A Key To Armageddon

– PETER FOOT –

An environmentally friendly book printed and bound in England by
www.printondemand-worldwide.com

Mixed Sources
Product group from well-managed forests, and other controlled sources
www.fsc.org Cert no. TT-COC-002641
© 1996 Forest Stewardship Council
FSC

PEFC Certified
This product is from sustainably managed forests and controlled sources
www.pefc.org
PEFC
PEFC/16-33-415

This book is made entirely of chain-of-custody materials

FastPrint Publishing

www.fast-print.net/store.php

A Key To Armageddon

ISBN 978-178035-260-2

First published 2011 by
FASTPRINT PUBLISHING
Peterborough, England.

Acknowledgements

I wish to acknowledge the assistance of my good friend Richard Riley in the editing of the rough drafts of this book, and thank him for his enduring patience and good humour in the many hours this has taken.

Dedication

This book is dedicated to all Royal Marines, past, present and future.

Chapter One

The night had been bitterly cold in this north-eastern corner of Afghanistan, not surprising as November heralded the approach of the Afghan winter. The Royal Marine Commandos' ambush position had been carefully chosen, with the five Marines just a few feet apart on the steep side of the boulder-strewn gully. Corporal Ron Bedwell, who was in charge of the small group of Marines, looked hopefully towards the east for signs of the approaching dawn; a glance at his watch suggested it would be soon. The Marines had been in this ambush position for five days and were due to be taken out by helicopter later that morning.

The track through the gully was one of the many routes into Afghanistan from the Pakistan border just five miles away. The Taliban received most of their supplies through these many narrow, rock-lined gullies, carried by small family groups anxious to make some money and gain favour. The Russians, who had tried and failed to defeat the Taliban, had mined many of these routes in a desperate attempt to break the supply chain. But over the years the mines had been exposed by wind and rain and now provided the essentials for the roadside bomb.

Ron Bedwell and his Marines breathed a sigh of relief as the first shafts of sunlight began to brighten the early morning sky. Almost at once they heard the sound of movement from the entrance to the gully. A minute passed; the sound of footsteps grew louder and then the travellers came into view. Instead of a gaggle of well-armed tribesmen it was a disappointment; just a man, his wife and one small boy riding on the back of the first of two well-loaded donkeys. Corporal Bedwell challenged the small family group, telling them to stand still. As they did so, he instructed three of his Marines to search the family group for any contraband.

As the Marines moved down to carry out their task, Corporal Bedwell instructed his signalman, Marine Arthur Jones, to contact base and inform them they would shortly be withdrawing from the ambush position, and to send the helicopter. The Corporal failed to see the father give a signal to his son, as he and Jones were now behind a large boulder. The son acted on his father's direction and, with a brief final glance at his mother, pressed something strapped to the donkey he was riding. There was a blinding flash as the explosives carried by the trusting donkey detonated. The sound echoed and re-echoed around the rock-covered hillsides as the dust and debris of the resulting explosion fell back to earth.

Corporal Bedwell and his signalman were stunned and deafened by the blast, but otherwise unhurt. Of the family, their two donkeys and the three Marines, no living trace remained. Arthur Jones, unable to hear the response to his message, resorted to sending his signal by Morse code, hoping the Army receivers at headquarters could read Morse. Cpl Bedwell walked out to where the family and his three Marines had been; it was like a scene from a horror movie. Body parts were everywhere; only the human and animal heads appeared to have survived in one piece from the explosive force. Only fragments of adhered clothing offered a clue to the identity of the remaining flesh-and-bone fragments.

Almost two hours had passed before the sound of the approaching helicopter grew near. It was a Chinook, so someone could obviously read Morse and understood the urgency of the message. Within ten minutes a still partially-deaf Cpl Bedwell was explaining to the Unit Intelligence Officer the sequence of events following the arrival of the Afghan family and their two donkeys. As they spoke, four Army medics were busy collecting the human remains and bagging them up to be sorted out later at base. Only the more obvious remains of the two donkeys were left to provide a meal for the land and air predators, which would arrive within minutes of the helicopter's departure. One of the medics handed a small metal case that had survived the explosion to Lieutenant Peter Chard, the Intelligence Officer. He decided to let the ordnance team examine it first back at base, in case it was booby-trapped.

With a final glance at the scene of their friends' deaths, the two remaining Marines and the recovery group made their way back to the Chinook. Very little had been said and the noisy Chinook prevented any further conversation on the return journey. The four plastic bags containing the body parts lay unattended, near the retracted ramp of the helicopter. Trained and dedicated men and women would be tasked to place in the three Royal Marines' coffins their remains, with some ballast to make up the correct body weight. When the coffins were flown home for burial by their families they would be sealed and would remain so. Sadly, their wives and parents would have no final sight of their much-loved son or husband.

On arriving at Camp Bastion, the two surviving Marines were seen by the Duty Medical Officer, Captain Clare Rogers. For those older service personnel who could remember the late Hattie Jacques, Clare Rogers was almost a perfect double. She dealt with all her patients in exactly the same way, be they a Private soldier or a Major General. As she dealt with the two Marines, their unnaturally loud voices suggested they both had a hearing problem, and the explanation they gave confirmed this. Fortunately it was not serious, and both Marines should recover their normal hearing after a good night's sleep.

Peter Chard carried the small metal box over to the Ordnance Section for them to check if the case contained explosives. The Sergeant there viewed the object with little concern but, just to make sure, he took it over to the dog compound. There he offered it to a small, excited spaniel, which showed not the slightest interest in the object but devotedly licked his hand. Back in the workshop the Sergeant carefully opened the metal box with a small screwdriver and showed its contents to the Royal Marine Lieutenant. Two computer memory sticks lay in its neatly-padded interior. Both men looked surprised; it was an unusual package for an Afghan family to carry.

Peter Chard took the package over to the Intelligence Office to show to his superior, Major Rob Furlong of the Army Intelligence Corps. The Major viewed the metal box and its contents with no great enthusiasm.

He sighed and said: "Well, Peter, I am not going to stick these in our computers to see what the memory sticks contain. It could be the mother and father of all viruses and shut our system down for months."

The two men discussed the memory keys for some time, finally deciding to pass on the problem to GCHQ at Cheltenham.

That evening Lieutenant Beverley Chambers was flying home to England because she had become pregnant. She had been instructed to deliver the small parcel she had been given by hand. It was not the responsibility of the package that occupied Beverley's mind. It was the delicate matter of explaining to her husband how she had become pregnant, when she had not seen him for all of six months. She had a good idea who was responsible; it was after a mess party just over three months ago, when the guests had been three American F16 pilots. It had been the dark-haired one with the Hollywood smile who had taken advantage of her drink-dulled wits, and screwed her up against the side of the Brigadier's Land Rover.

It was just after dark when the Royal Air Force transport aircraft started its flight to England, with a twelve-hour stopover in Cyprus. Forty hours later and a very weary Beverley Chambers arrived at GCHQ Cheltenham. She had already decided how to solve her problem by having a private abortion in a London clinic. She had been kept waiting at the reception area at GCHQ for thirty minutes when a rather officious female executive took her to one side and enquired why she was there. The package was passed over with the sealed letter, and then Beverley was on her way to London to arrange her more urgent business. The executive officer watched the young Lieutenant leave the building. "My word, she is in a hurry," she thought, wishing she could have spent more time with her. Tilly Longton was a very active lesbian, part of a coven at Cheltenham who revelled in the name of 'The Amazons'.

It was the next day when the letter and package reached the desk of one of the Chief Scientific Officers. He read the letter first before turning his attention to the metal box and the two memory sticks. He opened his staff file and looked at several names then, having decided on who would be best suited to evaluate the two computer keys, picked up the box and set off down the passage.

At Room 16 he stopped, paused for a second or so to take a deep breath, and walked in. The sour body smell of its occupant was equivalent to walking into a room full of unwashed rugby players' jock-straps. The Scientific Officer did not hang about. "Evaluate this for me, Frank," he said, before turning and making a rapid exit.

Frank Grimley smiled and broke wind, last night's 'Chicken Korma and rice with the Grimley special garlic sauce' adding to the fetid atmosphere in which he thrived. Grimley was a quite brilliant intelligence analyst; his work was his life, with little time for personal hygiene and female distractions. He looked at the two memory sticks and read the attached note. He had two computers in the room; one was his own, and both had the latest anti-virus programs installed. He placed one of the sticks into the computer's console and brought the computer to life.

To his frustration, three sentences in Arabic appeared - clearly, instructions on how to proceed further, with what secrets the key held. Frank picked up the phone and dialled an internal number. It was answered almost straight away, the voice saying: "Sam Yousof here."

Grimley told Yousof of the text in Arabic and asked him to come over and translate. "Have you washed recently?" enquired Sam Yousof, who was fastidious in matters of personal cleanliness. When Grimley made no reply to that question, Yousof suggested the text could be passed to him by computer link. In response, Grimley's expert fingers raced over the keys, and he received an almost instant acknowledgement.

A few minutes went by; Frank Grimley scratched his itching crotch, completely unaware of what he was doing. The phone rang.

"Sam here, Frank. The instruction relates to the information being suitable for the flight computer systems of Boeing 747 and 757 aircraft. Does that help?"

Grimley thanked him and replaced the phone, his brain already working on an explanation of what use the memory sticks would be. The most significant fact must be how the sticks or keys were obtained. Clearly, if they had been in Taliban hands, it would not be for a peaceful purpose. He moved the information on; it was now in 'Computer Speak'. Frank Grimley correctly assumed that on entering the program it would replace the one that had been previously entered. Frank now realised that to make sense of what the program contained he needed a qualified Boeing pilot to explain its meaning.

Frank Grimley phoned the Scientific Officer who had dumped the two memory sticks in his lap, and explained the problem. Determined to let Grimley sink or swim with this delicate task, the Officer told him to contact Bristol Airport and get some advice.

"But do not let the keys leave here," he ordered. This freedom of action was most unusual at GCHQ and Frank Grimley was determined to make the most of it. Within minutes he was speaking to the Operations Manager of one of the major airlines who, when realising he had been contacted by GCHQ Cheltenham, became most helpful.

Chapter Two

A recently serviced Boeing 747-8 aircraft slowly emerged from the maintenance depot at Heathrow and was being towed over to Gate 21. In three hours time it would begin its journey to New York with three hundred passengers and crew aboard. Two maintenance workers were still on the aircraft, one still checking the internal phones and the other an electronics expert at work on the Flight Deck. Ravi Pashto was deleting the last flight instructions from the flight deck computer so that the Captain and First Officer could enter the details of the flight they were about to make. He was, however, doing much more than that, as he inserted a memory stick into the computer and transferred the information it contained.

His task completed, Ravi Pasto removed the key and, making sure he was not observed, bowed his head and called on Allah to bless his work. By this time the aircraft was at Gate 21 and the two maintenance men climbed onto the towing vehicle for a ride back to their depot.

The Captain and First Officer were in their airline's Operations Room checking the paperwork concerning weather, passengers, freight and their fuel load. Captain Rex Masters was United Airlines senior Captain; a former American Air Force bomber pilot, he was widely experienced and respected. His First Officer was somewhat overdue for promotion to Captain, mainly because of his reputation with the ladies. Mark Longman had been blessed with the good looks of a film star, and had the nickname of 'Handsome'.

"Anything special we should worry about, Mister?" asked Rex Masters, as he glanced across at his First Officer, who was checking the freight manifest.

Longman did not reply immediately. Then, as he reached the list's end, he said: "No, Captain, it's all pretty innocent".

It may not have appeared so to any onlooker, but the two men could not stand the sight of each other. Both, however, were professionals and would conduct the coming flight as a team. Their briefing completed, the two men made their way to their aircraft to check that the ground crew had put sufficient fuel aboard and that the Senior Cabin Steward was happy with the food and drink allocated. There were two Air Stewards and eight Stewardesses on board for the flight; 'Handsome' Mark Longman had already slept with five of the girls and had his sights on the other three.

Longman completed his walk round the outside of the aircraft, checking that all the locking plates on the control surfaces had been removed and that the tyres on the landing wheels were in good order. As he entered the aircraft one of the girls he intended to pursue on this flight was leaning over one of the seats, adjusting its angle of comfort. Her blue uniform skirt stretched tight across her neat bottom, and Longman made a mental note to list her as target one. He made his way to the flight deck and for the next twenty minutes he and Rex Masters fed into the Flight Computer their flight details, including fuel poundage and alternative airfields.

A few minutes later the Senior Steward reported that they were ready to take on their passengers. Masters nodded and said: "OK, go to it, Joe."

Thirty minutes later they had embarked all their passengers. The engines were started one by one, beginning with the port and then starboard inboards, then the outboard engines in the same order. Permission was given to leave the gate and join the departing aircraft; they were fifth in line. Under their own power now they followed a Boeing 737, keeping the approved distance between aircraft. With the passengers and cabin crew all seated and belted up for take-off they were now third in line to the take-off runway.

Two minutes passed and, with the 737 now half-way down the runway, Captain Masters moved his 747 on to the runway threshold and waited for the control tower to give the final permission for take-off. It was a tradition with most airlines for the Captain to fly the aircraft on take-off and the First Officer to handle the subsequent landing. As permission was given for take-off, Rex Masters said: "Brakes off, Mister," and then opened the four throttle levers, causing the aircraft to begin its accelerating run to its take-off point. Longman, with his eyes fixed on the instruments, said: "Rotate" and Rex Masters eased back the control column and the massive 747 climbed effortlessly into the late afternoon sky.

A few minutes later the seatbelt sign was switched off and the cabin crew moved to check the comfort of their passengers. On the flight deck the Captain and First Officer were following the instructions from ground control on height and course to steer. Within the hour and in the growing darkness the last faint lights of the west coast of Ireland were left behind them. With the aircraft now at its approved altitude of thirty five thousand feet Captain Masters switched control of the aircraft to autopilot. The computer was now flying the aircraft in accordance with the programme the Captain and First Officer had chosen; or was it?

With a nod to his First Officer, Captain Rex Masters undid his seatbelt and left the flight deck to have his customary walk around his passengers; there were two Senators aboard who he knew. As he had left the flight deck closing the door, his First Officer had muttered "Stuffy old bastard", then let his mind wander over the prospect of adding the neatly-arsed Stella Asher to his 'have screwed list'. Masters spent ten minutes with the Senators then paused to speak to an ageing former 'western film star' who was accompanied by a stunning-looking girl young enough to have been his granddaughter. The girl was obviously not related, as the former 'star' had his hand active under her brightly-coloured skirt.

Masters stopped at the flight attendants' galley and was given a mug of coffee, strong and sweet, just how he liked it. Captain Masters dallied ten minutes at the galley, and then returned to the flight deck.

"Any problems, Mister?" the Captain enquired.

Longman shook his head and said: "It's a smooth ride so far, Captain."

Rex Masters grunted and then said: "Take a break Mister, but remember the cabin staff have plenty to do."

As Longman left the flight deck he was furious that the old bastard had warned him from socialising with the flight attendants.

The memory key or stick that Ravi Pasto had used to insert an aggressive program into the aircraft's computer was already doing its work. The aircraft was already jettisoning fuel from the aircraft's main tanks, completely unnoticed in the darkness. Normally, if this were to happen, warning lights would show on the flight deck's display panel and an alarm would sound. The sophisticated program that had been entered into the onboard computer by Ravi Pasto had negated this, and already the aircraft had insufficient fuel to reach the American coast. An hour later, with both pilots on the flight deck, the steady low noise of the engines changed slightly and it became obvious to both pilots that the outboard starboard engine was losing power.

Rex Masters grunted: "Controls to manual, Mister." First Officer Longman responded by switching off the flight computer. Immediately the flight deck instrument display panel became alive with flashing red warning lights and alarm buzzers. Captain Masters exclaimed "Jesus Christ" and switched off the warning sounds.

Both men stared open-mouthed at the display, realising the disaster it represented. Suddenly the outboard engines on both wings shut down and the Boeing 747 began to lose height. Both pilots knew their aircraft could fly on two engines, but not at the height they were at. But that was not their major concern; a lack of fuel for their still-functioning engines was.

"Get off a 'Mayday', Mark, and ask for all shipping in our area to be informed."

"Yes Captain," said Longman, well aware for the first time ever that Rex Masters had called him by his Christian name.

In response to the 747's 'Mayday' Newfoundland Air Traffic Control came back to them loud and clear, asking the extent of their emergency.

Masters turned to his First Officer: "Tell them we will be ditching in the next thirty minutes on this course, and are there any ships that can come to our assistance?"

Longman passed the message and was told "Wait on". At that moment the Chief Steward came onto the flight deck. Captain Masters quickly briefed him. "We are going to ditch, Joe; get your team together and prepare the passengers for a wet landing."

There were no questions from the Chief Steward, just a nod and a "Yes, Captain" and he left to try and control his staff and the two hundred and eighty-eight passengers.

Both of the remaining engines were now beginning to hunt for fuel. Captain Rex Masters knew that in just a matter of minutes he would have to put the aircraft into a glide angle to maintain enough speed to prevent his aircraft falling from the sky. Newfoundland Control came back to them.

"We have you on radar, Captain, and there are two ships ahead of you some twenty and thirty miles distant; keep us updated on your plans."

They were now at fifteen thousand feet and the Captain was forced to increase their angle of descent to keep the aircraft in the air as the remaining engines shut down.

Below in the Atlantic, the Captain of the Southern Venture listened to the conversation between Newfoundland and the Boeing 747. Sea conditions were poor, with south-westerly gusts of wind reaching almost forty knots - poor conditions for a survivable landing. Fifteen minutes later, with the aircraft now at five hundred feet, both pilots were struggling to keep the aircraft from stalling. Both men knew they needed to touch down in the sea with the nose slightly raised and the wings level, to make the crash-landing one that their passengers could escape from. With their landing-lights on, Mark Longman was tasked to report any sight of the sea below them. Suddenly he called out: "About a hundred feet to go, Captain."

Captain Rex Masters eased back on the control column to bring the nose of the aircraft up, but the Gods had decided otherwise. A sudden, forty-knot gust lifted the port wing, causing the tip of the starboard wing to catch the water first. The result was a total disaster; the aircraft cart-wheeled, ripping the starboard wing away and plunging most of the fuselage nose-first and upside down into the sea. The only survivors of the crash were in the rearmost seats of the aircraft's tail-section, which had broken away as the aircraft had settled. Seven passengers and one flight attendant were left struggling in the bitterly cold sea as the remainder of the massive aircraft disappeared from view.

Some twenty minutes later the Southern Venture arrived at the debris-strewn spot and recovered the eight persons; sadly, only five were still alive. The rescue ship stayed on the scene till daylight hoping to recover more victims of the crash, but without success. The one surviving crew member was Stewardess Stella Asher. Her evidence to the subsequent enquiry team was uncomplicated: "We simply ran out of fuel."

This information left the experienced team of former pilots scratching their heads, because the ground crew had witnessed a full load being pumped aboard, and no experienced pilot would cross the Atlantic with less.

Chapter Three

Peter Savage eased his six-foot, solid frame back behind his desk after his thirty-minute daily gym workout. Now freshly showered and looking the very image of a Royal Marine Officer, Savage started to wade through the pile of papers on his desk. Madge Ryan, his motherly secretary, came into the room and handed him a paper with a red-lined border; this indicated it required urgent attention.

"The boss wants you on this straight away, Peter. I'm to give the rest of your papers to the schoolboys to play with." They both laughed because, at thirty-one and a former Royal Marine, Major Savage was by far the only world-experienced man in the section. All the rest were selected University Graduates in their first or second year at MI6, all with potential but sadly lacking any experience of the real world.

After three tours in Afghanistan and two in Iraq, Major Peter Savage RM MC and Bar had decided to leave the Royal Marines and try something different. In the first week of his discharge leave, spent with his widowed mother in Horndean, Portsmouth, a letter had arrived with a job offer in London. The interview took place in a scruffy office block behind Victoria Station. To Savage's complete surprise, one of the three-man team conducting the interview was a Royal Marine Brigadier he had once served under. After a twenty-minute chat, Peter Savage had been offered employment with a Security Company better known as MI6.

That had been a year ago and Savage had taken to his new role like a duck to water. His work took him around the world meeting all sorts of people, not all of them friendly. Peter Savage was a romantic and, although he was successful with women in the physical sense, he had yet to meet the one he wished to spend his life with. Savage turned his attention to the red-bordered paper and sipped the tea Madge had brought him. The paper concerned the strange circumstances of a Boeing 747 running out of fuel half-way across the Atlantic, causing the deaths of almost three hundred people. The paper also suggested he went to GCHQ, Cheltenham, to begin his enquiries and contact a certain Mr Grimley there.

Peter Savage contacted Frank Grimley and arranged to see him that afternoon. Madge had already found out the time of a suitable train from London with a convenient connecting service to Cheltenham. It was just before three o'clock when the taxi dropped Savage outside the policed gates of GCHQ. Having produced his identity card, Peter Savage was kept waiting at the gate until an escort could take him to Grimley's place of work. In response to the MI6 officer's polite knock,

Frank Grimley quickly answered the door. Frank had put on a clean shirt for the occasion but had not gone as far as a good wash. As the door opened, Savage was met by an odour you could equate with a very-infrequently-cleaned toilet.

Frank Grimley shook Savage's outstretched hand and invited his guest to take a chair in the debris of his office. The MI6 officer took the lead by asking what had been discovered in the examination of the two memory keys from Afghanistan. Frank Grimley wiped his nose on the back of his hand and told of the assistance the senior airline pilot had provided.

"He told me that if the memory sticks were used in the aircraft's computer prior to the flight then, irrespective of the information the flight deck crew fed into the computer concerning their flight, the memory stick's program would take precedence."

"Does that mean that the flight deck crew's flight plan is disregarded?" asked Savage.

Grimley smiled and shook his head. "Oh no, that is the clever part; the input from the memory stick converts the flight crew's plan into what the inserter of the stick or key intends the fate of the aircraft to be."

"By discharging its fuel in flight, then?" said Savage.

"Could be," said Grimley. "In fact, any function the crew require the aircraft to perform could be compromised."

Savage thought for a second or so then asked: "In what way?"

Frank Grimley glanced at some notes he had made during his meeting with the airline pilot.

"Well, my expert advice was: landing-gear failing to deploy, flaps operating on one wing and not the other, rudder failing to respond and engines shutting down."

Peter Savage digested the information he had been given, and then asked: "But surely the flight deck crew only need to return the flight controls to manual and the problem is solved?"

"That is what I thought," said Grimley "But my pilot said the memory stick may have been programmed not to release control back to the flight deck."

"Did both of those memory sticks have the same program installed?" asked Savage.

"Yes," said Grimley "But stick two was to be used on an Airbus 380, and that has five hundred passengers."

As he returned to London that evening, now in possession of the memory sticks, or keys as some experts preferred to call them, Savage had a good deal on his mind. He knew that the recent crash of the Boeing 747 would be explained away by a suggestion that the aircraft's fuel had become contaminated, causing engine failure. The conversation between the crew of the doomed aircraft and Newfoundland Air Traffic Control was now classified and not for release. As far as the statement by the stewardess was concerned, the official spokesman stated: "Miss Asher was mistaken in her belief; she was not on the flight deck so she could not possibly know the state of the aircraft's fuel load."

On reaching London, Savage went first to his office and placed the memory keys in his safe. Then, though it was now nine o'clock, he had ordered a late meal in the staff restaurant and enjoyed the three-course meal with a glass of wine. His accommodation was in a block adjoining the MI6 building. It was quite comfortable and consisted of a bathroom, bedroom and small lounge, far superior to that in a Royal Marines Officers' Mess. Savage watched the late-night news. Two more deaths of British soldiers in Afghanistan from roadside bombs were the main items. He slept pretty well, awoke refreshed and had a shave and shower, then went for breakfast. He was playing in a badminton competition that evening so he gave a miss to his gym workout.

He was in his office by eight-thirty. Madge was already there with instructions from his section head; they were: "take the two items to the boffins then come and see me". Savage did just that, ignoring the scientific section's moans about their heavy workload. He knocked on his section head's door and was invited in. The former Royal Navy Captain was struggling to light his pipe. At last he succeeded and asked how helpful GCHQ had been. Savage gave a full account of his visit, leaving out the irrelevance of Frank Grimley's hygiene. Ralph Maitland listened carefully to what Savage had to say, and then sighed.

"It is all getting so complicated, Peter. Shortly, some electronic device will replace the pleasure of a good fart."

Twenty minutes later, with the next stage of the investigation decided, Savage went back to his office to have a coffee with Madge and to phone Special Branch. It was decided to bring in Special Branch because, if further enquiries led to a criminal prosecution, they were the people to handle it. As it was, MI6's role in this case was edging very near home-based MI5's territory even though the Boeing747 had crashed in mid-Atlantic. Peter Savage had now been informed that the aircraft had been brought to its departure gate by maintenance engineers after a major service. Clearly he needed to talk to these people, who would have been the last persons near the flight deck before the actual crew.

A senior Special Branch Officer put down the phone and considered the request from MI6 for one of his team to work with an agent investigating an air crash. Superintendent Jack Walsh knew MI6 could not at this stage have a suspect. If they did, he knew the probable police role would be to investigate a sudden death. He decided to nominate Sally Gunn for the task; she was a decent young girl, hard-working and totally reliable. Just to satisfy his selection he checked her file. She had been in the police for almost four years and had joined CID after two years of beat experience. Jack Walsh noticed a letter of resignation marked 'Withdrawn' in the file - clearly life in CID had not gone well. She, however, had stuck with the role she had been given and had been recommended for transfer to Special Branch on merit.

A few minutes later, Sally Gunn was knocking on the door of the 'Super's' office, curious at the abrupt summons to the boss's office.

"I am lending you to MI6 for a few days. Be on your guard that they don't involve you in any dirty tricks."

The Superintendent smiled. “Don't look so worried; it could be that their enquiries might lead to a prosecution, which is where you would become involved as the investigating officer.”

A few minutes later Sally was driving her aging Hyundai Atos towards the immaculate and imposing Secret Service Headquarters, hoping to be allowed into their massive, underground car-park. After producing her Warrant Card, she was kept in a ‘waiting bay’ for about ten minutes while her so-called assignment was checked. As she patiently waited, trying to remain calm, Sally considered the time she had spent in the police. She had to admit that the year she had spent in CID had been character-building. Her stomach churned as she recalled those most unhappy days when suicide seemed the preferred avenue of escape.

The first few days had set the scene for what was to follow. She was allocated to Inspector Roy Marshall’s team, which consisted of two Sergeants and eight Detective Constables. One of the Sergeants was female but she was more masculine than her male counterpart. Sally Gunn was one of two female Detectives in the eight-person team. Their male colleagues had little use for female officers, who were expected to make the tea or coffee, wash the cups and respond enthusiastically to a sweaty-handed grope on demand. Sally had responded to the first attempt by a male colleague to slip his hand up her skirt with a cracking slap across his face. The man responded by loudly saying: “Oh for fuck’s sake - we have been lumbered with another ‘dyke’.”

Sally's musings were interrupted by the security guard leaning into her car window and saying: "Someone is coming down, Miss, in a few minutes."

Sally remembered the unhappy weeks and months that followed in that CID hell. Her fellow female Constable suggested she let one of the lads screw her.

"I do; it keeps them happy and makes life tolerable, even if they do call me the section 'bike'."

After almost a year there Sally had put in a letter of resignation. She was summoned and asked why, so she told them. A few days later she was offered a transfer to Special Branch and grabbed it with both hands.

Someone tapped on the window. "Follow me, Miss; keep your speed down to five mph." The young man then rode off on a smart scooter with Sally Gunn close behind. The underground car-park was massive and policed by several security staff who demanded her car key, thereby preventing an unauthorised exit. The scooter-rider took her to the lift and escorted her to the third floor and an office marked as 'C1'. As she entered, a middle-aged woman invited her to sit down and said: "Major Savage will only keep you waiting a minute or so." A buzzer sounded and the secretary indicated she was to go in, saying: "Don't worry, my dear, he won't bite."

Chapter Four

Sally Gunn was pleasantly surprised as she entered the office to be greeted by a tall, well-built, smiling man who actually stood up with his hand outstretched to greet her.

"My name is Peter Savage. Do you wish to be addressed as Miss Gunn or Sally?"

"Oh," said a very surprised Detective Constable, "Sally will do just fine."

"In that case, please call me Peter," said Savage, pleased that his Special Branch helper seemed to be bright and enthusiastic.

A smiling Madge brought in two coffees, and then Peter Savage ran through the circumstances of the air crash. He also said that the aircraft had just been serviced and that the maintenance team had taken the 747 to the boarding gate at Heathrow just prior to the flight.

Sally Gunn had listened carefully to what Peter Savage had said.

"I take it that we now pay a visit to the maintenance team there?" she asked.

Savage nodded. "Yes, we will go over there straight away. They are not aware we are coming, so no one has had the opportunity to call in sick."

They made the journey in Savage's almost new Vauxhall Astra, not a vehicle that would stand out among London's flashier road-users. At the entrance to the row of maintenance hangars a security guard directed them to one used by United Airlines. With Sally Gunn at Peter Savage's right elbow, they went to the manager's office and introduced themselves. The manager, an American, was not overjoyed to see them, and tried to delay his cooperation by making a time-consuming phone call to his United Airlines Head Office.

Savage was used to this reluctance and quickly changed the manager's mind by saying: "I can and will suspend all flights made by your airline if you do not cooperate." "How can I help you?" said the manager, trying hard to smile. Over the next twenty minutes the manager produced the worksheets covering all the work carried out on the doomed aircraft.

"Are the people who carried out all the work on that aircraft here today?" asked Sally Gunn, who was completing a list of names from the worksheets.

Pausing to check his day-book, the manager said: "Yes, every one of them."

Peter Savage then asked: "Who worked on the electronics of the aircraft? Was there more than one person involved?"

"Oh no," said the manager, "Young Ravi Pashto is a genius on all those systems; his father was employed here before he retired. He was equally gifted, but on hydraulic systems."

Peter Savage turned to Sally Gunn. "I think we will start with young Pashto; do you agree?"

"Yes, Sir," said Sally, noticing with some amusement the look on the manager's face at such respect.

"Have you a room we can use for the interviews?" asked Peter Savage.

"Yes," replied the manager. "That one over there is free."

A few minutes later the smiling face of Ravi Pashto appeared at the office door. He was invited in and shook hands with both of his inquisitors, not knowing who they were. Peter Savage told him they were making enquiries about the loss of the 747 Series 8 aircraft which had disappeared over the Atlantic.

"I am told," said Savage "That you were still working on it when it was towed over to its departure gate."

For the first time young Pashto looked uncomfortable and looked at Peter Savage as though to anticipate the next question.

Sally Gunn took advantage of this and surprised him by saying: "What were you doing on the journey from the maintenance area to the departure gate?"

Pashto stammered his reply. "I forget, I can't remember."

"Was it to do with the flight deck computer?" asked Savage, for the first time real menace in his tone.

Ravi Pashto looked very uncomfortable and then replied as if in a trance. "I wiped the memory of the flight deck computer ready for the Captain and First Officer to enter their flight plan."

"How do you delete the previous program?" asked Savage. "Is it by pressing the delete button on the control panel or do you use another method?"

"I can use a memory-cleansing stick, which is quicker than using the deleting system on the control panel," said Pashto, now quite visibly nervous.

"And on the day in question?" asked Sally Gunn, sensing that Ravi Pashto was close to cracking. The young electronics engineer took several quick breaths before answering.

"I think I used the memory stick; yes, I definitely used the stick because we were at the boarding gate by now and I wanted a lift back on the towing buggy."

Savage gave the young engineer a long, hard stare, and then said: "Thank you Mr Pashto. That will be all for now, but just one thing. My colleague will accompany you to your workplace and take possession of the memory-cleansing stick that you used; we will return it to you when we have finished our enquiries."

Sally Gunn followed Ravi Pashto as he left the office and they made their way to his workplace. Not a word was said. Sally returned a few minutes later with the memory stick and handed it to Peter Savage. On the shank of the stick was a strip of adhesive tape; written on it was 'Ravi's'.

Both Peter Savage and Sally Gunn were satisfied that Ravi Pashto had played a role in the aircraft's fall from the sky, but they needed proof. As a matter of course they spent the next two hours talking to other members of the team that had serviced the 747. Then they made their way back to MI6 Headquarters. After a coffee in Peter Savage's office, Sally Gunn made her way back to Special Branch to find out what they had on Ravi Pashto. She had been told to return to MI6 the next day and had been given a pass allowing access to the complex and its car park. Peter Savage had sent the Pashto memory stick to the 'boffins' for expert analysis. A curious Madge enquired how he had got on with Miss Sally Gunn. Savage smiled and said: "I treated her as though she was your daughter, Madge."

Savage put Ravi Pashto's particulars into the MI6 computer. Nothing of note emerged. The Pashto family had settled in England after the war in 1945. Pashto's great grandfather had been an Officer in the Indian Army and later an instructor at Sandhurst Military Academy. That evening Peter Savage played badminton for an hour then, after showering, went for dinner. He had meat pie, with three vegetables and apple crumble with custard, then returned to his room and watched 'New Tricks' on his television. Savage had to admit Amanda Redman was impressive in her role and, unusually for a woman, made no attempt to disguise the scarring on her upper left arm. At eleven-thirty Savage took to his bed and enjoyed an almost dreamless sleep. He awoke at six-thirty, ready for almost anything.

Just before nine the next morning, Savage arrived at his office somewhat surprised to find Sally Gunn already there and in deep conversation with Madge. On Peter's desk was the Ravi Pashto memory stick and a report stating it contained a well-used 'deleting programme'. Sally Gunn had found only one thing on Ravi Pashto to interest Special Branch; a few months ago he had almost abandoned his Muslim faith and converted to that of a Roman Catholic. It would seem family pressure had convinced him to stay with Allah.

Peter Savage and Sally Gunn spent some time that morning discussing Ravi Pashto. What were his motives, and who was controlling him? The fact that he was a Muslim suggested a terrorist group, but which one? Madge brought in some coffee and a plate of biscuits; it was then that all the lights went out. Within five seconds the emergency generator had restored their power, but a glance out the window revealed the surrounding area and its office blocks were not so lucky. Savage's phone rang; it was his section head.

"Just heard from our friends at 'five' that the computer controlling London's electricity supply has selectively shut many essential services down."

One area of London that seemed to have been targeted was the huge city financial centres. They, of course, screamed louder than the emergency services which, of course, had their back-up systems. The Stock Exchange was forced to cease trading and certain Members of Parliament rushed to phone the Prime

Minister's office to complain. Peter Savage and Sally Gunn thought as one that, yet again, a computer was causing chaos and agreed that a visit to the control centre might give them a further lead in their investigation.

On arriving at Deptford, Peter Savage found his MI6 pass opened doors that a policeman would have required a warrant to open. By now the majority of the areas that had suffered a sudden loss of power had been relieved to find it restored. One of the last to be brought back to life was the Stock Exchange. The Duty Controller at Deptford was unapologetic to the angry voice of big business, telling him in no uncertain terms that people on life-support machines would always be their first priority. With the situation now almost back to normal, Savage and Sally Gunn shared a cup of tea with the Controller.

They at last were able to enquire what had caused the shut-down. The Controller lit a forbidden cigarette, took one puff and quickly put it out before replying.

"That is what I intend to find out. It was as though the computer had suddenly developed a mind of its own, completely diverting from its prepared program."

Savage and Sally Gunn exchanged glances, now most anxious to find out more.

"Who sets the computer's program? Is it done on a daily basis or on a longer term?" enquired Peter Savage.

"Oh, we might not change the program for months," said the Controller. "We route our energy, or electricity as you call it, through a system of sub-stations, avoiding overloads."

"I understand that," said Savage "But who changes the programs?"

The controller paused, and then said: "My two-man team, both computer wizards."

The Computer wizards always worked the two-till-ten shifts, which gave them the evening reduced-demand period to make adjustments to the system. On being told this, Savage said he and his colleague would like to meet them during the evening period. He added: "We would like our visit to be a surprise, can you arrange that?"

The Controller agreed and the pair made their way out of the building, pausing to look at a photo display of the staff at the security desk. Nothing was said until they were on their way back to the MI6 building, then Savage said: "Did you notice our two computer wizards on the photo board?"

Sally Gunn laughed. "Oh yes, they and our Ravi Pashto could be related."

It was almost seven forty-five when they returned to the Control Centre at Deptford, and within minutes were being introduced to the two men. The men explained that they had spent most of their time since coming on shift trying to explain the power shut-down.

"Have you found the answer yet?" enquired Peter Savage.

"Well," said one, and then looked at his colleague, who nodded his head. "We took a print-out of today's program and compared it to one we entered some weeks ago." He then laid two print-outs on the table and in red ink circled the changes between the two.

Sally Gunn leaned forward to examine the changes, and asked: "How did that occur then, gentlemen, by mirrors?"

The older of the two men smiled and said: "No, Miss, someone changed the program."

Chapter Five

In the north-western corner of Pakistan is the large town of Jinnah. It possesses two large schools, both having a reputation for excellence. Most of the employment the town provides is with a large electrical manufacturer which produces a vast number of kitchen appliances. In recent months the firm has opened another factory nearby, mainly concerned with assembling mobile phones from parts shipped in from China. In another part of the newer building the factory's most talented young workers are involved in producing computer programmes and working on anti-virus devices. The young and talented manager of this department is Asif Abbass, and he is given a free hand in selecting his workers.

The Abbass family is one of the largest in the area. Asif has three older brothers and two younger sisters. Asif's father has five brothers, all married with large, growing families, some now living on the border with Afghanistan. Over the last five years the Abbass family had become quite wealthy. The money had not come without risk and their clients never accepted any excuse for failure or disloyalty. Asif was twenty-two and as yet unmarried; his father, who was a lustful man, was concerned that his son may have had leanings towards his own sex but his wife assured him this was not so. She knew that her son was involved with a young widow and another woman who, though attractive, was old enough to be his mother.

Later that afternoon Asif was meeting one of his clients at a lay-by about two miles outside of the town. Asif made the journey on his old Vespa scooter. Usually the client would arrive by car; it would be in immaculate condition and almost new. Asif had a wait of less than five minutes for his client and, when the car arrived, he was invited to sit in the back. The man's companion on the rear seat was Chinese like her master. Su Yang was very attractive and wore a very short skirt to emphasis her excellent legs. At a nod from her companion the girl obediently moved into the front of the car next to the driver.

"I have a new project for you to programme, my young friend," he said, then produced a folder which he opened and showed to Asif. "This is a print-out of the process of activating part of the American missile shield and making them ready to fire retaliatory missiles."

Asif swallowed hard, very aware of what he was getting into. If the client recognised Asif's concern he ignored it, and went on: "I want you to complete the firing process, but to launch the missiles into this map reference area, which is in the Bering Sea fifty miles east of Kamchatka."

"Is it to be programmed by memory stick?" enquired an anxious Asif.

The client smiled. "Yes, produce two keys; we have two agents with access to the North American Defence Complex."

Asif Abbass went to leave, but the client said "Wait." He then beckoned his female companion to rejoin them on the rear seat. She did so, exposing a great deal of her upper legs in the process.

"You like?" said the client. Asif nodded.

"Then screw her now on this seat any way you like, it's your bonus." Asif Abbass did just that, the girl's reluctant participation adding to the excitement, as did the client by urging him on and the driver watching eagerly in the rear-view mirror. When Asif had finished, his place between the girl's legs was immediately eagerly taken by the client. Asif Abbass picked up the folder and exited the car, not wishing to play gooseberry. As he rode back to his family home he wondered if the driver would be permitted to complete the girl's humiliation.

The following day Asif tasked one of his best workers to copy the complete program from the folder onto the computer. They then guarded the program by adding a series of passwords based on the Christian names of early American Presidents. Asif took the trouble to check the six-figure map reference he had been given and was relieved to find it did relate to an area fifty miles off the coast of Kamchatka. That evening before he went home

Asif ran through the program again, making one or two small adjustments before signing it off as the finished article. When Asif arrived home his mother was about to serve dinner. His father asked him what plans he had for the rest of the evening. Asif, well aware from what his mother had told him that his father doubted his sexuality, replied that he was going to play table tennis with a very nice young man he had met. His father's face darkened to a purple shade until Asif's mother told her husband that his son was teasing him. Humour in the family was restored and they all sat and enjoyed the food their mother had cooked, Asif's mother giving her husband a conspirator's wink.

In Rawalpindi, one of Pakistan's top intelligence agents sorted through the accumulation of messages that were heaped on his desk. His efficient secretary was sorely missed and her return could not come soon enough. Jali Mohammed gave a sigh of relief as he eventually sorted the wheat from the chaff and selected four messages for his immediate attention. One concerned a Chinese businessman who was part-owner of an electrical product factory in Jinnah. The other was from Peter Savage in England concerning the sabotage of computer programs by the introduction of damaging viruses or contradictory information. The Chinese businessman was known to him as Lee Peng. They had met twice before during enquiries into the laundering of vast sums of money and the purchase of land for development.

Jali lived for his work and, since the death of his wife in childbirth, it was a way of escape from the pain of her loss. The other two messages were requests from the

American and British military to overfly the Pakistan border with their drone aircraft. Jali Mohammed knew that whether he gave permission or not the flights would take place, so he tore the requests into shreds and binned them. He thought that the message from Peter Savage and the factory part-owned by the Chinese man known as Lee Peng could be connected. His secretary was due back the next day having been away to comfort her mother and arrange her father's funeral, so Jali decided to visit Jinnah straight away.

His vehicle was a Honda off-roader and ideal for some of Pakistan's appalling roads. The journey to Jinnah took three hours. He arrived just after midday and a surprised Asif Abbass took him to the factory's canteen for some lunch. Afterwards they went to Asif's office and Jali Mohammed enquired if Lee Peng had visited recently.

"Yes," said Asif "I saw him yesterday. He wants us to produce a new model of mobile phone, very sophisticated and certain to be a big hit in the west."

"Anything else?" asked Jali.

"Oh, he allowed me to screw his female companion in the back of his car as a bonus," said Asif.

"Is that the one with the superb legs and a skirt up to her armpits?" said a smiling Jali.

"Yes," said Asif "A very attractive girl; her name is Su Yang."

Jali Mohammed asked for a tour of the factory before he left and took a particular interest in the newer part, where the most talented of the staff worked. Asif would have preferred his visitor not to have seen the computer-programming part, but he had little choice. The security man asked what these workers were doing.

"Oh, they are compiling anti-virus programmes; it is an important part of our work," explained a cautious Asif. Shortly after, Jali began his journey back Rawalpindi; he had his suspicions about what went on at this factory, especially the newer part, and his young guide was clearly very apprehensive at his interest.

In one of Karachi's finest hotels Lee Peng and his companion, Su Yang, had just returned to their room after a couple of hours by the hotel pool. With his business in Pakistan completed, Lee Peng would fly to Singapore the next day. The international cartel of which he was a member had a meeting scheduled there in the next few days. The timing of these meetings was a closely-kept secret and the participators were barely given more than an hour's notice of the event. Su Yang was in the shower, so Peng decided to join her. When they were at the pool she had attracted many a male eye with her excellent figure. Lee Peng was always turned on by the attention that his property attracted.

Su Yang was unaware that her master had joined her in the shower until two large hands captured her breasts and she felt his erection against her bottom. His voice growled his instructions and she complied, despite her head banging painfully against the shower wall as she was bent almost double by the vigour of his passion. Later, as they dressed for dinner, he presented her with a beautiful and clearly very expensive pearl necklace.

"I do not find it easy to express my affection for you, my dear, but perhaps this gift will make up for my lack of words."

Su Yang was moved, and took her master's hand and kissed it before saying: "I am yours, Master, now and forever."

The next morning they took a noon flight to Singapore, travelling first class of course. They arrived there in the late evening during a torrential downpour. Not a drop of rain touched them as they were whisked by covered transport to the arrival desk, then by the hotel limousine to the luxury of the Continental Hotel. At breakfast the next morning Lee Peng noticed several of the Cartel members already there. Acknowledgement was made by the briefest of nods. Su Yang noticed her master's action but she was more interested in their female companions, realising the standard of the competition.

It was late afternoon when the Cartel meeting took place. All of the members' female companions (only one or two were wives) were taken on a tour of the Island. A strange choice was a visit to the 'Tiger' brewery, where they were given champagne and entertained by four male strippers, who all possessed huge private parts. The ice was broken and the conversation flowed freely. When they arrived back at the hotel the meeting had just ended and it was time to dress for dinner. Breaking with tradition, they all dined at one large table then enjoyed the evening's entertainment, which included a world-famous recording artist.

The following day most of the Cartel had booked out of the hotel by mid-morning. Lee Peng and the delicious Su Yang were the last to leave. Lee Peng had purchased, about a year ago, a beautifully restored villa in northern Malaya not far from the old tin-mining town of Ipoh. Tambun House had a very mixed history. It was originally owned by the Sultan of Perak as a hunting lodge then, during the Japanese occupation in World War Two, it was used as a rest house for officers. After the war and during the Malayan Emergency a succession of British Units was based there. Su Yang loved the place and often said that if she woke during the night she always felt secure, as though guarded by the military occupants of the past.

Chapter Six

The USS Dallas was a guided-missile destroyer and had only recently joined the American Fifth Fleet, based in Bahrain. It was at present in the Arabian Sea, where it had been dispatched by its Admiral to work up to an operational standard. Captain Greg Meadows was determined that his ship would be the most efficient one in the Fifth Fleet and his officers and crew had already experienced the rough edge of their Captain's tongue. Greg Meadows had worked his way up from the lower deck in twelve years, a considerable feat in any of the world's navies. His one weakness was his belief that he was despised by his fellow officers over his success. This was not true; in fact, they admired him and wanted to be good friends, but Meadows was convinced he was resented.

When Captain Meadows had arrived with the Dallas at Bahrain, he had been greeted warmly by Admiral Zac Polter, who had said: "Replenish your ship, son, give your crew a couple of days leave then take the Dallas into the Arabian Sea and work your crew up until you are satisfied they are ready for anything." Captain Meadows had done just that, testing the crew's reaction to 'Action Stations', 'Damage Control' situations, and even 'Prepare to repel boarders'. By day three of their working-up exercise the crew was deliberately slowing down its reaction times and morale was falling through the ship's bottom.

The ship's Executive Officer was Lieutenant Commander Peter Taylor, a decent man who, despite his privileged upbringing and coming from a wealthy family, liked his Captain and wanted to help. That evening he went to the Captain's cabin and asked if he could speak frankly. A surprised Greg Meadows said "Yes"", confident he could handle anything. Taylor did not mince his words and for almost five minutes said his piece. He concluded by saying: "All of your officers and crew are ready to follow you anywhere, Sir, but they feel that you do not trust them; do that, Sir, and you will have the most efficient ship and crew in the US Navy."

After Taylor had left his cabin Greg Meadows sat trying to contain his anger, then he gave serious thought to what his Executive Officer had said. Could he be right?

After a far from restful night he made his way to the Bridge, ignoring his steward's question about breakfast. On the Bridge he acknowledged the salutes and the "Good morning, Sir." Then he surprised the most junior officer on the Bridge by asking: "What shall we do this morning, Mister Webber?"

Webber's face was a picture, but when his Captain quietly asked again, "Well, Mister?", he sprang to life, saying: "I would like to rig the ship for towing a damaged and burning ship aft, Sir."

Meadows nodded and said: "Then go to it, Mister."

Aided by a very experienced Chief Petty Officer, Mr Webber quickly assembled a towing rig and fire hoses on the stern of the Dallas. It had taken just fifteen minutes. The Captain was impressed, not only with the ship's most junior officer, but also the enthusiasm of the crew. Conscious of what his Executive Officer had said the night before, Captain Meadows congratulated Webber and his team then went to his cabin for breakfast. That afternoon he allowed the crew to swim over the side with two guard boats in the water for safety. They spent two more days at sea then returned to Bahrain, now a more efficient ship and, just as important, a contented crew.

A week later the Admiral sent three of his destroyers to sea to test-fire at a towed target a new batch of Sea Snake missiles. The new version of the Sea Snake was vastly more effective than the old Exocet of the Falklands war. The Sea Snake, when launched from a ship or aircraft, could approach its target just a few feet above the sea and below the guard of radar. The three destroyers in line-ahead formation and followed by a fast patrol-boat towing the target, reached the test-firing area in sea conditions of flat calm. The towed target was just a twenty-foot square of white canvas and, if the missile passed through the canvas, it would detonate about a hundred yards on the other side. The fuse on the Sea

Snake's nose was set to explode with a slight penetration delay so as to achieve maximum damage to its victim.

The towing vessel and its target were now in position and the senior Captain of the three destroyers would lead and fire two Sea Snakes. In line ahead the USS Boston prepared to fire on the first run, with the USS Dallas and the USS Memphis following him and watching with great interest. The USS Boston fired his first, then second, missile. They were blindingly fast and just a few feet above the sea. The target was about one thousand yards away. Missile one missed the target by fifty yards and began a slow but complete turn back towards its firing point, the USS Boston.

Missile Two missed its target by the same amount and also began a returning course, but this time its target seemed to be the Dallas. What was happening on the bridge of the Boston could only be guessed at, but on the Dallas the shock and surprise lasted but a second or so. Lieutenant Commander Taylor turned to his Captain with an urgent "Sir!" Meadows nodded and shouted: "Weapons Officer, target the missile with the bow Phalanx." With the approaching missile now less than five hundred yards away the Phalanx opened fire. With a noise like the ripping of some stout canvas cloth the Phalanx fired three two-second bursts. Some three hundred 20mm explosive rounds blocked the missile's path and it detonated a hundred yards from the ship.

The officers on the Bridge of the Dallas now turned their attention to the fate of the USS Boston. It was a mass of fire and smoke and listing badly to starboard.

Captain Meadows called out to Junior Lieutenant Webber.

"Mr Webber, you have experience in these matters; rig the ship for towing a burning ship aft and get the hoses into play." The Boston was now almost dead in the water and the Dallas eased slowly alongside. If the Boston's arsenal of explosives were to explode not one but two ships would be lost. On impact the missile had struck the Boston just below the Bridge and this had ceased to exist, which meant the Captain and senior crew were probably dead.

Captain Meadows sent his Executive Officer onto the Boston with a small first-aid party to rescue whom he could. An hour later, with the badly-damaged ship now in tow and its major fires out, the bow of the Dallas was like an emergency ward. The USS Memphis sent a medical team aboard Captain Meadows' ship to help and several helicopters from the Fleet's headquarters in Bahrain were on their way to extract the more-badly injured. An hour or so later the last of the two helicopters departed with the more-seriously injured. The Dallas, towing the shattered wreck of the USS Boston, began its slow journey back to Bahrain with the USS Memphis in attendance like an anxious mother hen.

To make their return journey more difficult a storm with attendant rough seas slowed them down still more but, after a difficult twelve hours, the conditions improved. Thirty-six hours later they edged the damaged USS Boston alongside the jetty, where a shore party took over. Captain Greg Meadows noticed the Admiral and some of his staff closely observing the damaged vessel.

Meadows eased his command to its berthing-point ahead of the Boston and watched as his crew secured the ship to the dockside. His Executive Officer, Lieutenant Commander Peter Taylor, suddenly appeared on the bridge, saying: “Captain, the Admiral is coming aboard.”

Without waiting for a traditional ‘Honour’ side party to greet him, the Admiral and two of his aides hurried to the bridge of the Dallas, anxious for a full account of the tragedy. Acknowledging Greg Meadows salute, Admiral Zac Polter quickly said: “Let’s go to your cabin, Captain, and you can give me the full story.” The Captain’s steward quickly brought some coffee to the cabin and handed the cups round to the guests, then hurried from the cabin. The Admiral took a couple of sips then, turning to Greg Meadows, said: “Ok son, let’s have chapter and verse.”

In less than ten minutes Greg Meadows described the sequence of events leading up to and then including the first missile hitting the Boston and the second destroyed on its way to the Dallas. The Admiral digested the information then spoke quickly to one of his aides, who rushed from the cabin. He then turned to Captain Meadows and enquired: “How come you were able to get your Phalanx deployed and shoot down the threat to your ship and the Boston did not?” Greg Meadows realised the implications of the question. “It was a question of time, Admiral. I and my crew had at least ten seconds - double that of the Captain of the USS Boston - to save our ship.

The Admiral said nothing for a few seconds, and then got to his feet. “I want you and your Exec to dine with me tonight at eight, dress informal; we need to get to the bottom of this, and fast."

Greg Meadows and Peter Taylor, in spotless khaki drill, arrived at the Admiral's shore headquarters fifteen minutes early. A steward escorted them to the Admiral's dining room to join the other guests; including the Admiral there were six in total. It was all very relaxed and friendly, and not until dinner had been completed did the real business of the evening begin.

Admiral Polter fixed Greg Meadows with a friendly smile and said: "Run through the complete sequence of events, starting with the Boston firing its two missiles please, Captain, and leave nothing out."

Greg Meadows did just that, the account finishing when they docked back in Bahrain and the Admiral came aboard. The Admiral looked at Lt Cdr Peter Taylor and enquired: "Anything to add to that, Mister?"

Taylor shook his head. "No, Sir; that is exactly what happened."

The Admiral looked around at his guests. "Right, gentlemen, you are all specialists in the missile field; ask your questions."

For the next hour Greg Meadows and his Executive Officer answered, to the best of their ability, a variety of questions.

The Sea Snake is a new missile; had you witnessed them being fired before?

Were your missiles and those on the Boston and Memphis from the same manufacturing consignment?

Did you actually see the first rogue missile impact on the Boston?

In answering question three, only Lt Cdr Taylor actually witnessed the first missile's impact. A variety of questions followed their answers and then the Admiral called time. The other guests departed, leaving the Admiral, Greg Meadows and Peter Taylor alone. American ships are 'dry'; in other words, no alcohol is allowed aboard, but shore bases are different and the Admiral's quarters were technically 'ashore'. Zac Polter opened a drawer and removed a bottle of scotch and poured out three double measures. They toasted the dead and injured on the Boston, the President of the United States and then the Navy.

It was now time to go. The Admiral shook them both by the hand and said: "You, Captain, and your crew, did just fine, but I would expect nothing less. Now both of you get some sleep; you look as though you need it."

The two men slowly walked back to where their ship was berthed; they glanced at the USS Boston tied up some hundred feet astern of the Dallas. It was smoke-blackened and its upper Bridge-deck was a mass of twisted metal. Nothing was said as they boarded their ship, the Captain leading the way up the gangway as tradition demands. As Greg Meadows prepared for bed he forced himself to complete his daily journal while the events of the day were still fresh. At last it was completed and, with his conscience now clear, he took to his bed. It was not a dreamless night and he found himself battling with a host of flying insects, all poisonous of course.

Then they disappeared and he slipped into a deep and comfortable sleep.

Chapter Seven

The head of MI6 replaced the phone and sighed, thinking what gutless bastards these politicians were. She had to admit this new lot were just as bad as the previous bunch. Dame Helen Butters was in her sixth year as head of 'Six', having been in the organisation for fifteen years. 'Marge' was how her friends called her and was the nickname she was known by in the security business. Though she was a highly-skilled intelligence analyst and spoke German, French and Spanish fluently, one of her friends had to explain to her why she was called 'Marge'.

She gave the recent phone call from the Defence Minister some thought. It concerned the missile mishap involving the USS Boston which had cost fifteen lives. It was obvious that the rapid advance of technology had opened the door to a new form of sabotage involving a sophisticated type of computer virus. She quickly glanced at the interim report Peter Savage had submitted on the two memory keys recovered in Afghanistan and the downing of the American Airlines 747. The phone call from the Defence Minister was all part of a politician's survival strategy - inform the intelligence services and it's their problem when the shit hits the fan.

Peter Savage had spent the weekend at the home of a former Royal Marine Officer who had suffered severe injuries to his legs as the result of a roadside bomb in Afghanistan. On the Friday evening after a meal he had helped Bob McKay to bed with the aid of Bob's very attractive wife, Mary. He then had spent the rest of the evening reading the rough notes of Bob's first attempt at writing a book; it was on the life of Rudolf Hess. In the notes, Bob McKay had freely admitted that his story followed the detention of Hess on landing by plane near the Duke of Hamilton's estate in May 1941. What followed has been wreathed in secrecy to the present day, with the suggestion that a member of the British Royal Family was involved in arranging a peace treaty with Adolf Hitler, with Hess acting as Hitler's envoy.

As Savage read the notes, it was clear that his friend's attempt to obtain information about the arrival of Hess and his subsequent imprisonment was obstructed at every stage. This only added to the mystery; and why was he imprisoned in the Tower of London? Prince George, who was the Duke of Kent at the time and later died in an air crash, was suggested as the link between Hess and the British Government. But with Churchill now in charge, no peace deal was possible. At the end of the war Hess was tried at Nuremburg and given a life sentence which, on reflection, seemed excessive. He was then kept in Spandau Prison in solitary confinement until August 1987, where he was said to have committed suicide.

Being aged ninety-three at the time and suffering from severe arthritis, the ability to succeed in taking his life by hanging only adds to the mystery.

Savage had noticed in the notes that a former medical assistant at Spandau had alleged Hess had been killed by British Agents because the Russians were arguing for Hess to be released on humanitarian grounds. There was also a note from a Government department saying that all papers in the case of Rudolf Hess were not available for public scrutiny for a further fifty years. Bob McKay had scribbled a note in the margin referring to the Government memo, saying "is this to avoid any embarrassment to the present Royal Family?"

On Monday morning, as usual, Madge Ryan was in the office ahead of Peter Savage. When he came in she commented that he looked as though he had missed some sleep. Savage gave a weary smile and admitted that he had been helping an old friend and his wife over the weekend. Madge gave a disbelieving sniff and said Marge wanted to see him at eleven with the Section Head. Savage took the memory-key folder and their work on the 747 out of his desk drawer and read through the contents once again. Sally Gunn was at Heathrow making a list of all maintenance workers with access to aircraft flight decks. She had so many names that she was working on a system to eliminate a good few of these possible suspects by checking their previous employment.

At eleven o'clock Peter Savage and his Section Head were outside the Head of MI6's office. An aide showed them in, then served coffee and took a seat, his notepad at the ready. The head of MI6 spoke: "The country's leaders are at an early stage of filling their pants over this latest incident with the United States warship; they fear it is our turn next." Helen Butters looked at Savage then his Section Head for some response. "Well, Madam," said Savage, "They could be right. To prevent this type of incident you need to be ahead of the game and we are still catching up." His section head nodded his agreement. "I think Peter's right. We have alerted all three Services on the need to be extra vigilant in all live-firing exercises, and Peter has an acquaintance in Pakistan who has offered help."

As they left the Head of MI6's office Savage asked his Section Head if he was familiar with the file on Rudolf Hess and the restriction imposed on it. The former Royal Navy Captain gave Savage a curious look and said: "What on earth has that to do with our present problem? That old bastard has been dead over twenty years." Savage was embarrassed and decided to come clean and explained about his old Royal Marine colleague's injuries, and his attempt to write a book on Hess but present it as fiction. Ralph Maitland scratched his chin then wagged his finger under Peter Savage's nose and, smiling, said: "All countries have secrets to hide, Peter, and we are no different. If you start sniffing around sensitive and restricted intelligence your career in 'Six' could be a short one."

Across London Sally Gunn was in the final stages of reducing her list of suspects. Ravi Pashto was, without doubt, top of the list, but they had no firm evidence against him. The fact that he was one of the last people on the flight deck of the 747 was suspicious, but nothing more. She decided to speak to Peter Savage and made her way to the MI6 headquarters, comfortable that at least she had somewhere to park her car. She liked Peter Savage; he seemed a decent man and was one of the first male colleagues she had worked with who had made no attempt to get into her knickers. Of course, he could be gay but she thought not - being in the Royal Marines made this unlikely but not impossible.

When she arrived at Peter's office Madge gave her a welcoming smile and said: "Go in, dear, he is expecting you; security phoned to say you were on your way up." Savage was on his knees under his desk retrieving some papers which had slipped to the floor. He finally emerged with his face just inches from the hem of Sally Gunn's short skirt. "I am so sorry," said Savage, "Please sit down while I sort these papers out." Sally did so, thinking to herself that this guy must be straight, going to such lengths to have a glance up her kilt. Peter Savage finally resumed his seat and enquired if she was any nearer to identifying a suspect worth pursuing.

Sally shook her head. "I have looked at all the personnel having access to the aircraft's flight deck and Mr Ravi Pashto has to be the main suspect, but he is as clean as a whistle and apparently he does not even fart in bed." "How many mobiles does he have?" asked Savage. "Two," said Sally. "Both have a trace on them, but so far they have only carried conversations of his social life." "Why two mobiles?" asked Savage. Sally Gunn laughed. "He seems to keep one for contacting a colleague's wife, who he is clearly shagging when this friend is on nights."

Elsewhere in London, two men who had recently arrived on a business trip from Pakistan were also discussing Ravi Pashto. "We understand our trigger man is being watched; he is their only suspect in the 747 operation, after all." The other man thoughtfully stirred his coffee before speaking. "Time for disposal, then?" His colleague nodded. "It's arranged for this evening; the Underground during the rush-hour can be a very dangerous place." The man subject to all this attention was hard at work in one of Heathrow's maintenance hangars, replacing a faulty electronic guidance system on a Boeing 737. Ravi had plans for this evening involving the wife of his maintenance opposite number. Tara's husband was on nights so the pair would go to the cinema, then have a meal and spend the rest of the night pounding the life out of Tara's marital bed.

At the end of his shift Ravi Pashto collected his motor-bike and made a swift journey to his flat in Brixton. Ravi had found the motor-bike was the ideal form of transport to beat London's rush-hour traffic. He quickly showered and changed into his favourite light-blue suit, then walked the hundred yards or so to the Underground station. The platforms were crowded and he made his way to one servicing the Circle Line. A rumbling distant sound heralded the arrival of his train and the packed mass of travellers edged nearer to the platform's edge. As the train emerged from the tunnel with the customary blast of warm air, a hand propelled Ravi Pashto forward and into the path of the still fast-moving train.

Incredibly, only half a dozen people who were close by had witnessed the tragedy and it was several minutes before they could convince the station staff of what had occurred. The police and fire brigade finally arrived and, with the power to the live rail switched off, the search for the victim began. It took an hour for Ravi's remains to be recovered. He could only be identified by his Heathrow Airport pass.

A frustrated Tara waited almost an hour for her lover to appear and then angrily made her way home, determined to now end the arrangement. The next morning when her husband returned from his night shift, she was made aware of the sudden death of his friend and colleague. Her reaction surprised her husband, as she fainted at the news but, being a trusting and loving husband, he put this down to her caring and considerate nature concerned at the loss of their friend.

Both Sally Gunn and Peter Savage were not over-surprised at their chief suspect's sudden death. To them it merely confirmed their suspicions but, of course, the trail was now cold. If Ravi Pashtu's killer should ever be traced, it surely would have been a contract killing sealed with the handing over of a wad of used readies. The funeral of their chief suspect took place in accordance with Muslim custom the next day. Both Peter Savage and Sally Gunn had made sure no legal obstacle would cause any delay, because they were anxious to see who attended the ceremony. Every mourner was photographed secretly by Special Branch and a dossier compiled on each one - a rare occasion when the two rival services worked together.

The next day Peter's Section Head suggested that he should go and see his contact in Pakistan, and take Sally

Gunn with him. Sally was the more excited of the two at the prospect of some foreign travel, and the chance of finding some lead there that could advance their enquiry. Two business-class seats on Pakistan's National Airline made the long flight quite comfortable. As their seats extended into a comfortable resting position, Savage had to admit that Sally Gunn had excellent legs.

Chapter Eight

Some one hundred miles west of the island of Madeira, the hunter-killer submarine HMS Talisman was travelling south at about twenty knots and at a depth of three hundred feet. The Talisman was a Trafalgar Class submarine and carried a combination of lethal hardware comprising of Tomahawk cruise missiles and Tigerfish torpedoes. Talisman had a crew of one hundred and thirty, twelve of whom were officers. Her power was provided by a Rolls Royce nuclear reactor which generated the energy to power her two GEC steam turbines, capable of giving Talisman a maximum speed of thirty-two knots. A few days previously they had waited at Gibraltar, hoping to be deployed with their sister submarine HMS Triumph in action against Colonel Gaddafi's Libya, authorised by UN Security Resolution 1973.

To their disappointment they had been instructed to proceed to the South Atlantic, taking their turn to provide an underwater guard on the Falkland Islands.

Their tour of duty would be of two months, sealed in this steel canister, breathing air scrubbed clean of the deadly carbon dioxide, and without feeling the warmth of the sun on their faces. Drinking water for the crew is provided by the desalination plant and the submarine's rubbish, including body waste, is processed through a digester. It is then discharged at night as micro-particles. Submariners are a special breed and quite rightly subjected to a vigorous selection process, with emphasis placed on personality and mental strength.

Commander Mark Seymour-Watson was the Captain of the Talisman; he was married and he and his former-actress wife, Grace, had three children. Like most British vessels of war, when they are designed the accommodation for the crew is only considered when the engines, fuel tanks, magazines and weaponry are on-plan. At this point the crew's accommodation is then fitted in to the remaining space available. On submarines space is even more at a premium, with 'hot–bedding' the normal practice. The recent announcement by a Ministry of Defence spokesman that females would be permitted to serve on submarines was at first greeted with disbelief, and then with eager anticipation by submariners.

Modern ships of war have systems totally dependent on the sophisticated technology that state-of-the-art computers can provide. The Control Room of a submarine looks almost like the flight deck of a Star Wars spacecraft. The submarine's crew are trained in the use of this modern technology, but they all are aware that if something can go wrong, it probably will at sometime or other. All submarines have a Medical Officer who is also a trained surgeon; he has to assist him a small medical team, all highly trained.

On most submarines, in an attempt to save water, only the medical team and the cooks are permitted a daily shower; everyone else has to settle for an infrequent wash-down. If for some reason the submarine surfaces as part of an exercise, the Captain may allow a limited number of crew to take in the rare, fresh air. Strangely enough, this isn't as popular as one may think, as the change of air can bring on bouts of vomiting.

Ten days later the submarine had just passed the island of Ascension and was heading for its patrol area at twenty-five knots. The Officer of the Watch had called the Captain to the Control Room, as one of the two compasses there was showing a reading different from the other. It is common practice in most warships for the Captain to have a bulkhead-mounted compass in his cabin. Seymour-Watson sent a junior officer to his small cabin to check what reading that gave. It tallied with one of the two in the Control Room, so a technician was instructed to check the fault in the other.

Petty Officer George Fenwick spent two hours trying to locate the fault, decided the problem lay with the submarine's computer system, and passed the problem on to Chief Petty Officer Jim Simpson. Chief Simpson was an acknowledged computer expert, and Commander Seymour-Watson considered himself fortunate to have him aboard. An hour later a worried-looking Chief Petty Officer Simpson presented himself at the Captain's small cabin. "Sorry to bother you, Sir, but I think we may have a problem with the submarine's main computer."

"What kind of problem, Chief? Is it serious or a minor matter?" enquired the Captain.

Simpson scratched his head. "It's difficult to say, Sir. We have one of the best anti-virus systems fitted. They even use it on the Space Shuttle, but something is in there that just isn't right."

"What do you suggest, Chief? We are already double-checking the compass bearings."

"Leave it with me, Sir, I have some very bright lads in my team and we will work on it non-stop."

With that the Chief left. Commander Mark Seymour-Watson took his operation orders from his safe and read through them. He had little option other than to continue his patrol, as no other submarine was available.

Chief Petty Officer Simpson and his team worked on the computer fault until just before midnight, without success. They then decided to get some rest and try again the next day. Simpson briefed the oncoming Officer of the Watch on their lack of progress. Lieutenant Alan Roberts was also the submarine's First Lieutenant and, apart from watch-keeping duties, was responsible to the Captain for the discipline of the crew. At the half-way point during the Middle Watch, at 0200hrs, one of the watch-keepers drew the Lieutenant's attention to the panel displaying the submarine's computerised activities. "We are discharging waste from the digester, Sir." Lt Roberts glanced at the display, not concerned because this was a routine always carried out at night.

As the officer moved away from the display panel, the watch-keeper called out: "But Sir!" The officer sighed and said: "What now?"

"This can't be right, Sir. The desalination plant is taking on water as well. That should not happen."

Realising the implication of this act, the Lieutenant quickly ordered: "Go to manual control and switch off the desalination pumps." Despite the man's frantic attempts to go to manual, the computer refused to surrender control. Meanwhile, hundreds of gallons of polluted water from the digester were being sucked into the submarine's desalination system. "Cut the power and go to the emergency system," ordered the First Lieutenant, adding: "And get Chief Simpson up here at the double."

Mark Seymour-Watson, who had been asleep in his small cabin, woke instantly, realising something was wrong as the emergency lights came on and the submarine's fans cut out. Half-dressed he hurried to the Control Room, where his First Lieutenant quickly briefed him on the problem. "Is the fresh water system compromised?" he asked.

"Yes Sir, we were unable to shut the pumps down. The computer refused to allow us to go to manual control."

A weary and very worried-looking Chief Petty Officer Simpson asked to speak. "Yes Chief, please give us your assessment of this problem," said the Captain.

"In private, please Sir, away from the lads," suggested the Chief. The Captain nodded and moved to the small chart-room. "It seems to me, Sir, that our main computer has been got at. It would be dangerous to give it any role in the running of the sub's systems. If we are to continue this patrol it will have to be on manual."

Within fifteen minutes the submarine was almost back to normal, without any important and helpful assistance from the main computer system. The nuclear reactor and steam turbines were unaffected, having their own computer with an obligatory back-up. Navigation was not a problem, because at night they could float a receiver to the surface and check their position by satellite. Drinking and cooking water was their main problem; washing of their bodies was a luxury they could do without. The Captain sent off a flash signal to the Admiralty, stating their position and enquiring if any nearby shipping could supply bottled water.

An hour later a signal was received. "Proceed as ordered - a frigate is on its way from the Falkland Islands with bottled water. Should meet you within twenty-four hours." Over the next six hours the Engineering Officer and his team flushed out the desalination tanks several times. They could now be used to flush the submarine's toilets, but nothing else until they had been properly cleaned in dry dock. All the submarine's weapon systems could still be used, but not as efficiently as with the main computer system. Meanwhile Chief Petty Officer Simpson had, with the Captain's permission, made further attempts to try and locate the fault in the main computer.

In London that morning, the plight of the Talisman was to remain a closely-guarded secret; the Admiral in charge of all submarine operations insisted on this. One of his staff, a Commander, asked if they should tell the Minister; after all he was responsible to Parliament for the Royal Navy. The look the Admiral gave his subordinate was lethal. "We must keep this to ourselves, is that clearly understood? Tell a fucking politician and it will be all over the papers and television by this evening."

Meanwhile, in the South Atlantic, the Captain of the frigate HMS Decoy was busy confiscating all the bottled water in Port Stanley and filling any suitable container with the precious liquid from the island's sweet spring-water sites.

It was nearer forty hours than twenty-four when the Decoy was able to transfer its precious cargo to the now very thirsty submarine. Once the transfer was completed, the Talisman resumed its course to the Falklands and its patrol area. The Captain and his Catering Officer worked out a sensible water-rationing system, which meant they could make each re-supply of water last two weeks. Chief Petty Officer Jim Simpson worked on the rogue computer for hours on end, and finally decided it was not a simple fault. The computer's program had been tampered with, and the foul water discharge from the digester system and the subsequent intake to the desalination plant had been jointly programmed to occur on a specific date.

As the Captain listened to Chief Simpson's explanation, the full significance of this sabotage sunk in: were other submarines or ships at risk? He and his Communications Officer spent the next hour composing a coded message to the Admiralty, alerting them to the problem. On the Medical Officer's advice, the working temperature in the submarine was lowered a few degrees, on the basis that a cooler working environment reduced the need for the body to take on water. The submarine's crew were not greatly inconvenienced by the temperature change, and just slipped on a sweater.

The temperature at the Admiralty on receipt and decoding of the Talisman's signal went in the opposite direction, and the matter was discussed at the highest level. All Naval Bases involving the fitting, repairing or replacing of computer systems were alerted to rigorously check their security. With the Libyan Operation under way, no chances could be taken with missiles and smart bombs, the majority of them being computer-guided. After some discussion the warning was also passed to the Royal Air Force, but the problems incurred by HMS Talisman were not mentioned; after all, this was a Naval matter, and not to be sniggered over by any of the other Services. Despite the wish to keep this 'in–house' and the assumption that Naval Officers do not leak, their Civil Servants do. And by five o'clock, the Chief of Naval Operations had been summoned to Number 10 Downing Street to explain himself.

Chapter Nine

Sally Gunn had spent most of the day in her room at the Pegasus Hotel, recovering from the longish flight from London. Savage, who rarely suffered from any form of travel lag, spent his day going through some files he was being allowed to read at the local police CID department. Jali Mohammed had promised to see them the next morning at about ten in their hotel. Most of the paperwork Savage had been given referred to computer fraud, with many of the new mobile phones passing on their owners' account numbers to a third person, dialling the phone's secret secondary number. Clearly the shops selling these attractive new model mobiles had to be part of the fraud or were astonishing naive.

Savage returned to the hotel and went out to a poolside bar for a cold drink. He immediately spotted Sally stretched out on a poolside lounger, looking very attractive in a powder-blue, brief two-piece swimming ensemble. A large, florid-faced man standing next to Savage at the bar said: "My mate's about to move in on that piece, and he seldom fails to score." "Really," said Savage, and waited for the expert to perform. On cue, a tall 'Hollywood type', tanned and probably brainless, casually strode up to Sally's lounger and spoke to her. Sally Gunn turned to look at her admirer and took off her sunglasses for a better look. She then said just two words; the second was certainly "Off". The man took a pace back and hurried away, clearly disappointed. His florid companion at the bar muttered "fucking lesbian", and buried his face in his glass.

Peter Savage made his way over to Sally, who sat up and moved to allow him to share the lounger. It was unnecessary, as a smiling pool attendant slipped a spare one quickly in place. "I've just been propositioned," said Sally. "Really?" said Savage, "Good-looking boy?" Sally grimaced. "A real flash bastard. He had used a couple of cans of fake tan on himself, and it looked as if he had inserted a banana straight out of the fridge into his very brief swimwear; really pathetic." "Yes," said Savage, "I stood next to his mate at the bar. When you did not wet yourself in excitement at his offer, you were immediately labelled as a lesbian."

They stayed by the pool for an hour, and Savage told her about the fraud being carried out under the guise of the new mobile phone. They then went to their separate rooms to prepare for dinner. Later, in the dining room, they spotted Sally's attempted seducer seated at a distant table with a far-from-young, rather plump woman. He had obviously made an impression as the woman was stroking his hand and arm and displaying possession. His companion was at an adjacent table with his wife, who was clearly talking his head off. In an effort to find some peace from this marital harassment, the man was seeking solace in a large glass of strong spirit.

After their meal Sally persuaded Peter Savage to take to the dance floor. The resident band was running through a selection of classic dance numbers very much to Savage's taste. Sally fitted neatly into the Savage frame, and in very short order he became aware of the soft contours of her body as they moved in time to the dance rhythm. Savage began to feel aroused by Sally's close contact and decided it would be a good time to make their plans for the next morning, when Jali Mohammed would join them. If Sally was disappointed she did not show it, and drew Peter Savage's attention to the 'Hollywood type', who was dancing very close to his conquest with both of his hands clutching the cheeks of the woman's massive behind.

The next morning Savage had finished breakfast when Sally finally joined him. She quickly dealt with her orange juice and dispatched a pair of kippers and two cups of coffee. She had just poured herself a third cup when Jali Mohammed joined them. He clearly found Sally Gunn very much to his liking and, to Peter Savage's amusement, gave her the full treatment of his soft brown eyes and persuasive tongue. Jali took them both in his 4x4 to a police research unit, which was dealing with computer fraud; it proved to be a fascinating morning. A young police Inspector showed them some of the complicated scams that the international crook was using to gain access to account numbers and make withdrawals using passwords thought to be secure.

They had lunch in the police canteen, and then they were taken to the record section and shown photographs of suspects and those recently convicted of computer fraud. One file caused Sally to burst out laughing. Pictured on the front was her would-be seducer of yesterday, looking very 'Hollywood' in a white suit, fake tan and flashing blue eyes. On opening the file his florid-faced companion and his formidable wife were pictured; 'Hollywood' was actually their son. They were Australians and were well-known in the Far East for preying on vulnerable women of all ages. 'Hollywood' would bed them and pass to his father the woman's passport and credit cards for copying while the woman was asleep. The 'romance' might last just a few days, or if she was very wealthy a couple of weeks, until they had cleaned out her credit card accounts.

Having passed to Jali and his colleagues the whereabouts of this family of fraudsters, Sally and Peter Savage returned to their hotel. The next day they were going to the local prison, where a very clever computer expert was negotiating a sentence reduction in return for some startling information involving the Chinese military. They spent an hour around the pool, enjoying a tray of English tea and some delicious cakes. 'Hollywood' and his plump companion were about ten yards away, the woman's hand resting in a possessive manner on the top of her seducer's leg. She leaned across and kissed her companion's cheek and whispered a request. It was not too enthusiastically received, but her insistence was clear.

Sally smiled at Savage as a fairly weary 'Hollywood' and his companion made their way up to the bedrooms, the woman squeezing her lover's right buttock as if anticipating the delights yet to come.

Savage laughed as the pair disappeared and turning to Sally said: "I almost feel sorry for our young friend, having to service that eager lady about four times a day. You could say he earns his money."

Sally gave him a quizzical look. "Not the job for you then, Boss?"

Savage sighed. "No, my dear. I'm still an old-fashioned romantic - we could be a dying breed."

After breakfast the next day, Jali Mohammed took them to the local prison, where they could observe through a glass partition the jailed computer expert's plea for a reduction in sentence. Jali also told them that the 'Hollywood' family was at present being just placed under observation. The reason Jali gave was that a security company was carrying out its own reverse sting operation and, astonishingly, the plump woman was one of their top agents. Sally Gunn thought it was hilarious and commented that the plump lady clearly enjoyed her work.

At the prison they were given refreshment and then comfortably seated to witness the prisoner present his case for a reduction in sentence; he, of course, was unaware he was being observed. The prisoner was a citizen of Hong Kong, and a product of a mixed marriage between a European man and a Chinese woman. Thankfully the distaste felt towards the children of these marriages is almost a thing of the past, and expressions such as 'Eurasian trash' and 'Chi Chi' are now rarely used. The prisoner, smartly dressed for the occasion, was in his early fifties, and spoke clearly as he acknowledged his name and date of birth. His interrogator wasted little time on pleasantries, abruptly stating: "Well, what have you got to offer? And don't waste my fucking time with bullshit. I want some meat, not just the smell of it cooking."

The prisoner took a deep breath, and then spoke. "I worked for the Chinese Security Service checking applications for licences to trade in Hong Kong and mainland China." He paused and sipped some water. "After a year or so I was transferred to a security section tasked with providing government computer departments with anti-virus systems." He took another sip of water and removed his jacket. "I was praised for my work and then given a promotion and transferred to the military intelligence section. There, with three others, I was tasked with developing a computer program to infiltrate, confuse and insert new target codes into the United States Defence facility at the Pentagon."

His interrogator lit a cigarette and offered one to the prisoner, enquiring: "And how successful were you?"

"In July last year we completely disrupted for twenty hours the North American Air Defence Shield. We informed our friends in Russia that, should they wish, they could nuke America from coast to coast."

"Is that all you have?" enquired the interrogator.

"I have more," said the prisoner, "But I want something in return. I want four of my five-year sentence classed as time served. After all, I did not kill anyone, just touched some rather curious young boys and girls with my erect and rather large prick."

The interrogator pressed a button and two guards appeared, and were instructed to return the prisoner to his cell, search him and put him back in prison clothes. Jali turned to Peter Savage and Sally Gunn. "What do you think of that, then?" "Let the dirty bastard rot in his cell," offered Sally. Savage shook his head. "No, we need to get everything he knows. Offer him in writing what he wants signed by the Justice Minister. Then when he spills the beans, release him and arrange a tragic accident." Sally gazed at Peter Savage in admiration. Jali laughed. "Hardly the actions of a British gentleman, Peter. I am surprised you can sleep at night."

They were treated to lunch in the Prison Officers' Mess, Sally Gunn attracting a good deal of attention from both staff and waiters alike. They spent an hour with Jali and the Prison Governor, who had old-fashioned ideas on the application of justice. His ideas of punishments were simple: hang all killers, remove the right hand of all thieves and castrate all sex offenders.

On return to their hotel they both changed into swim-wear and went down to the pool. They quickly spotted the plump, eager lady and 'Hollywood' on adjoining sun-loungers. 'Hollywood's' fake tan had almost faded; he looked washed out, drawn and desperate for sleep. The plump enquiry agent was still pursuing her man and was whispering suggestions in his ear and lightly touching the top of his leg, with little response.

Back at the prison, several phone calls had been made to the Justice Ministry, and by early evening a deal of sorts had been arranged. In the meantime, the Governor was to return the prisoner's status to that of a remand prisoner with its attendant privileges. On being informed of this, the prisoner asked if he could order a meal from an outside restaurant, allowed under the Remand Prisoner Rules providing the prisoner paid. A local Chinese restaurant was given the order and the meal was delivered by taxi. It was examined by the Prison Guards and passed to the prisoner for him to enjoy. An hour later the prisoner rang his cell bell for attention. It was not immediately responded to and, when it was, the prisoner was found writhing in agony on the cell floor.

The Medical Officer was sent for and he ordered the prisoner to be taken to the Prison Hospital for treatment. By eleven that evening the man was dead with suspected cyanide poisoning, later confirmed by autopsy. The Chinese Restaurant was cleared of blame but the taxi driver had disappeared, along with all that useful information. It was obvious that the Chinese Secret Service had arranged the special meal or, at least, one particular ingredient.

Frustrated, Sally Gunn and Peter Savage flew back to London the following day. On the same flight was the plump enquiry agent, looking very satisfied with the result of her hard work.

Chapter Ten

At a forward American Air Base in the north-east of Afghanistan, two heavily-armed drone Predator aircraft were being made ready for flight. The beauty of these war machines was the versatility of their control. They could be programmed to fly to a specific location and take out a building or vehicle located by satellite. As an alternative they could be controlled from a forward base, or by someone in a concealed position not far from the intended target. Despite the drone's highly-accurate guidance system mistakes had occurred, suggesting (but never being acknowledged) that the Taliban possessed some system of blocking or distorting the drone's computer brain.

United States Army Captain Willard Henry checked his map and aerial photographs of the targets his two drones were intended to destroy. His task was to guide them to a certain location where two groups of Special Forces would take over the final control. Henry felt a little uneasy about giving over final control to these groups. Twice in the last three months, this type of operation had gone badly wrong in the final stage, with the wrong locations being destroyed. Having completed his final checks the Predator drones took to the air, flying about two hundred yards apart. Henry's role now was to fly them to their rendezvous position.

One of the three-man Special Forces teams had been hiding on the rock-strewn slope overlooking the small collection of dwellings for the last twelve hours. The night had been bitterly cold. Sgt Jeb Stone was fighting a losing battle in a desperate attempt to control his need for a crap. His two companions could see the humour of the situation as finally their Sergeant just managed to ease down his trousers in time as he lost control. A spluttering fart broke the early morning silence and a dog began to bark, causing the three Americans to raise their weapons. A minute or so later all was now quiet again; even the putrid stench of the Sergeant's bowel movement had thankfully faded. The presence of Predator 'Alpha' revealed itself on the Sergeant's miniaturised control panel. He quickly pressed two buttons to take over control of the drone and its weaponry.

A minute or so passed and the drone was now flying a circle at about a thousand feet just two miles away. The Sergeant made a final check and whispered to one of his men: "Light up the target, Billy Boy." Billy Boy switched on his laser beam, bathing the large, stone-built house in a light only the drone's sophisticated weapon system could identify. A green light briefly glowed on the Sergeant's panel and he quickly pressed the red firing-button; death was on its way. The eyes of the three-man team were now fixed on their target, the rough stone building which was a shelter for several senior Taliban commanders.

Both missiles struck within a split second of one another, their impact making the ground shudder. As the smoke and dust finally cleared, people rushed out of the collection of rough buildings that was their village. They gazed around them, then saw the site of the explosion about two hundred yards away on the rocky slope. A pack of the village dogs was already examining the scene, drawn by the scent of blood and scraps of flesh littering the site. Predator 'Alpha', its task complete, returned to the control of Captain Willard Henry who, of course, was unaware of the disaster. A few minutes later the other drone, which had been more successful, joined up with its companion and began the journey back to their airfield.

Three days were to pass before the full extent of the tragedy came to light. A helicopter took a group of Special Forces to the scene, where they found the group of rough dwellings deserted. The site of detonation by the drone's missiles was, of course, where the three-man team had lain hidden. All that now remained was some shredded clothing and some pieces of bone thought to be human; the local animal scavengers had removed anything edible. The bone pieces were quickly collected and would be returned to the United States, spread out in the three returning heroes' sealed coffins. It became clear that, if the weapons of the three men had survived the explosions, they would now be in the hands of the Taliban.

After an hour or so the team returned to the helicopter, and made the return journey back to their base. Later, with their unit's Intelligence Officer present and with mugs filled with comforting coffee, they began to speculate on what had gone wrong. Captain Willard Henry, who was attending this debrief, informed them that having reached the rendezvous position, Predator 'Alpha' had been taken under the control of the Special Forces team. Once the team had discharged the drone's missiles it would then automatically return to his control, and indeed had done so. He was asked if he had noticed anything wrong with the drone's behaviour at the time, and he replied "No."

One of the youngest Sergeants in the group enquired if it were possible for the laser beam directed at the target to be reflected back towards the laser-beam holder by a glass window or something similar on the target. There was silence as this thought was considered by the Intelligence Officer and Captain Henry. A phone call was made to the Battalion's Weapons Officer, who promised to ring them back; clearly he was not sure. An hour later the Weapons Officer rang them back to say he had been in touch with the manufacturers of both the missile and the laser projector, and had been informed this was not possible; a reflection lacked the intensity of the original beam.

After another hour the meeting broke up; just one thing had been decided: a thorough check was to be carried out on the batch of missiles that could be at fault. Also, the laser beam projectors were to be retested for any faults. Another week passed, with the Special Forces Group being very wary of its sophisticated weaponry, and then there was a breakthrough. A young weapon technician found that, on three of the thirty missiles he had checked, the computerised guidance system which homed on a laser beam had been reversed.

This information was devastating because it meant the missile would home on the laser projector and not the illuminated target. It now had to be considered whether this was a tragic manufacturing accident, or someone had deliberately reversed the guidance system in these missiles. In the United States the missiles' manufacturer was a respected company with a long history of providing all types of weaponry for America's armed forces. The Chairman of Thorn Military Systems was a personal friend of the current and last two Presidents. The phone call from the Defense Department took most of the colour from his face. George Butterworth summoned his loyal secretary from the outer office. "Get Harry Satchman here straight away, Mary; it's urgent and top priority." She was back within a minute or so, dreading what she had to say. "Mr Satchman is about to leave for the golf course, Sir; he says can it wait."

Butterworth grabbed the phone from his desk and punched out the security office number. A voice said: "Security here, Sir, how can I help?" "Get that prick Satchman up to my office pronto, in cuffs if need be," growled the Chairman. Two minutes later a breathless Harry Satchman was in the Chairman's office, dreading what was to come. "You did all the costing for the Predator Drone Missile System, Harry?" enquired a pale-faced George Butterworth, his voice quiet but full of menace. A surprised and somewhat relieved Harry

Satchman agreed that he had. "And where did we obtain the computerised guidance system? I know that was expensive, according to the Samworth bid."

A smug expression replaced the previous anxiety on Satchman's face. "Oh, I gave the Samworth prototype to a firm in Taiwan. They produced something similar for a third of the Samworth price." George Butterworth suddenly slumped back in his chair, his mouth open gasping for breath, his hands clutching at his chest. Mary, his secretary, rang the factory medical section and they immediately called an ambulance and rushed to give assistance. Two hours later George Butterworth was stable in a local private hospital and issuing a stream of instructions for Mary to record. "Don't let that bastard Satchman see this, Mary. Give it to my son and tell him he acts with my full authority."

Harry Satchman had decided to resign his position with Thorn Military Systems, not just over the Taiwan contract but because he was having an 'affaire' with the Chairman's young, second wife. When George Butterworth had sent for him that morning, he had been convinced that the old bastard had found out and was about to dismiss him. Meanwhile George Butterworth's son, Mark, was carrying out his father's instruction to the letter. He had ordered all the Taiwan-produced guidance systems to be destroyed and had phoned the Director of Samworth asking for them to supply urgently a batch of their systems under a long-term contract.

Thorn Military Systems surprisingly escaped criticism over the faulty technology, but the American Freemasonry is as forgiving of its members as fellow

European branches are. Not a word of the faulty guidance systems reached the media, being blocked under the guise of National Security. However, the American and British Intelligence sections in Afghanistan did exchange the information, and provoked a rapid check in all British armouries of any printed-circuit missile-detonation systems.

In London, Peter Savage found his file on rogue computer systems was growing larger by the day and prompted serious speculation about the use of a system that so very few really understood. The Defence Ministry in London was trying to explain to its members in the House of Commons that it had the matter under control, when clearly it did not. The Defence Secretary turned to his Senior Civil Servant, seeking advice. The Civil Servant, realising his Minister was looking for a 'fall guy' to take the blame, suggested an urgent review should take place on all missile contracts. "How much time will that buy us?" enquired the worried Defence Minister. "We could spin it out for at least six months, then release a heavily-censored report blaming National Security for the lack of information released," replied the very wary Permanent Secretary.

The Prime Minister was now becoming more deeply involved in Libya than he had ever intended when he decided to step out onto the world's stage. What concerned him even more was that he had ignored the advice of his far more experienced Foreign Secretary on this leap into the dark. He was also not best pleased when he was made aware that his Defence Secretary was trying to dodge the blame for some faulty equipment installed in some of the missile guidance systems. He decided that, when the Cabinet met the next day, there would be blood on the walls, and it would not be his.

Chapter Eleven

The beautiful island of Penang lies off the north-west coast of Malaysia. It has been described as "a pearl set in a silver sea", and few people would disagree with that description. It had been the home of Elizabeth Ascott-Gallop for the last twenty years. Five years earlier her husband, John, had died. He had been an astute businessman and had founded and developed a very successful brewery company, producing Pearl Lager. On her husband's death, Betty AG, as she was known to her friends, decided to sell the company, having decided on another way of occupying her time. The Singapore Tiger Brewery showed immediate interest, as Pearl Lager was popular throughout the region. Betty AG turned down two offers then accepted the third and, overnight, became an extremely wealthy woman.

She decided to purchase a property on the Island's east coast at a place called Sandycroft. During the Malayan Emergency in the Nineteen-Fifties, it had been a British Army rest centre run by NAAFI. Within six months she had transformed the place into a Finishing School for Young Ladies, and then spent two months selecting her staff. She eventually selected three married teachers and a tall, very intimidating Physical Training Instructor. All were female and heterosexual with the exception of the PTI, who was clearly lesbian but in a stable relationship. Two cooks and three cleaners completed her staffing arrangements, and now she advertised her new school throughout the region.

The ages of the girls would be between fourteen and seventeen and they could stay for one, two, or three years. The fees were based on what the parents could afford and, with a maximum of sixty girls, up to ten could be accepted with just a token payment. The object of the school was to produce sensible young ladies who could find a ready career in modern business at a junior executive level. Six months after opening the Ascott-Gallop Academy for Young Ladies, all the places were taken and they had a reserve list. They had also included drama and music classes in the evenings to take up some of their pupils' free time.

The young ladies attracted considerable attention from the local young males and it became necessary to employ some security patrols, especially at night. Three retired local policemen fitted nicely into the system and the problem of 'peeping toms' disappeared.

Betty AG was now in her early fifties and, many years ago before she met her husband, she had worked at GCHQ Cheltenham as an Intelligence Analyst. In the security business you are seldom completely forgotten, and Betty AG was not totally surprised to receive a letter asking her if she could find time to speak to someone from British Intelligence. More curious than really interested, she agreed and an appointment was made for her to meet a certain Mr Peter Savage at the Sheridan Hotel for dinner.

Peter Savage had spent a couple of days in Singapore talking to one of the Department's local agents who had contacts working for a Far East Division of Marconi International. He had been instructed to spend a day or so in Penang and meet up with the Principal of a Girls' Finishing School, who had once been well-regarded at GCHQ Cheltenham. Savage assumed he was to make contact in case any of the girls, on finding work in London, might be persuaded to pass information to the Department about the bank or business that employed them. He decided to spend just one night at the Sheridan Hotel, meet the contact, and then return to Kuala Lumpur and take a flight to London, where his In-tray was piling up.

He reached the Sheridan Hotel mid-morning and booked in for just the one night, stating that he had a dinner guest that evening. "No problem, Sir," said the very attractive receptionist, "I will arrange everything." Savage had lunch and spent the afternoon sitting out on his room's balcony, watching the antics of several very attractive girls who seemed to be in a competition as to who could display the most flesh. Despite the amount of flesh on display, Savage fell asleep for an hour and finally awoke, bathed in sweat from the hot sun; the girls had gone. After showering and changing, Savage read the Straits Times, and made his way down to reception to meet his guest.

He informed the receptionist that he was waiting to meet Mrs Ascott-Gallop. “Oh, I know the lady, Sir. If you wish to have a pre-dinner drink, I will direct Mrs Ascott-Gallop to you when she arrives.” Savage decided to have a long version of the gin and tonic, with plenty of ice and tonic; it was invigorating. A voice suddenly said: “Mr Savage?” Savage leapt to his feet to greet his guest, offering her a drink. Betty AG looked at his glass and said: “I'll have one of those, if I may.” Twenty minutes later they made their way into the restaurant and were shown to their table. The meal was excellent and they disposed of a bottle of white wine together.

At the coffee stage, Betty AG turned to Savage and said: “What does the Department want with me? It seems a lifetime since I was at GCHQ.”

Savage sipped his coffee. “Well, you know the score. Once you work for Intelligence, they never really forget you, and I suppose in your new work they think you could be of use.”

Betty AG laughed. “Oh, for God's sake! I run a girls’ finishing school. On leaving the oldest would only be about twenty, much too young for your games.”

“We would keep an eye on them and eventually they could be of use,” said Savage. “Don't dismiss this out of hand. Any young girl coming to London is at risk from the local trash - we would prevent that.”

"I'll give it some thought," said Betty AG. "But I would like a brandy before I go.”

The one brandy became two and, before the pair parted for the night, Savage accepted an invitation to visit the Academy the next day. This meant extending his stay by one night at the hotel, and rebooking his flights from Penang to Kuala Lumpur and onwards to London. The attractive and efficient receptionist did all this for him, and Savage regretted that he was not staying longer so he could show his appreciation. The next morning after breakfast, he took a taxi to the Academy, and spent the rest of the morning being shown round. He stayed for lunch and was given the opportunity to speak to the older girls, who were likely to spend the minimum time there. At three o'clock he said his goodbyes, expressed his thanks for the tour, and returned to the hotel.

There were two messages awaiting him; one was confirmation of his new flights. The other was a query from his Section Head at MI6, suggesting he asked Elizabeth Ascott-Gallop if she knew a lady by the name of Carol Andrews-Dixon, who lived in Cheyne Court, an exclusive part of London. It was later that evening when Savage managed to phone his lunchtime host. He began by thanking her for the most interesting tour of her Academy and for the lunch. He then asked if she knew a lady by the name of Carol Andrews, or Carol Andrews-Dixon. There was a long silence after the question, and for a moment Savage thought the phone had gone dead. At last Elizabeth Ascott-Gallop replied: "My God, Mr Savage, that is a name from the past."

She then went on to say that she and Carol had been friends for years after spending a term together at Westonbirt Girls' School. "The last time I heard from Carol was when my husband died - she told me she was running a club of sorts for gentlemen in London." As Savage was about to reply Elizabeth burst out laughing, adding: "To be absolutely frank, Mr Savage, I think Carol was running a rather high-class brothel with a very exclusive client list." After the phone call had ended, Savage began to wonder why his boss had asked that particular question. The next morning he checked out of the hotel and took a taxi to Penang Airport to start his journey home. At Kuala Lumpur there was a slight delay, and then he boarded his London plane, which happened to be from Singapore Airlines.

He managed to sleep some of the way, spending the rest of the time reading a collection of newspapers that the airline provided. It was almost midnight when his flight arrived at Heathrow some twenty minutes early. An hour later he was making a cup of tea in his quarters, before taking to his bed; the next day was Bank Holiday so he could have a lie-in. The news of the death of Osama Bin Laden in Pakistan filled all the television and radio stations the next day. Peter Savage, being a former Royal Marine, was eager to find out more, and contacted his MI6 Headquarters for an update on the latest information.

They, of course, had a more detailed version than the news organisation that was already spicing up its stories, with speculation that Osama died a coward's death, hiding behind his wife. Savage did not believe that for an instant; whatever your views on this terrorist leader were, he certainly was not a coward. Another piece of information caused Savage to raise an eye in query - there were no American casualties. This had to suggest that, contrary to expectations, this man was not defended by armed guards. As more information was revealed this was confirmed, and further information revealed that Osama

Bin Laden was unarmed but resisted capture and was shot dead in the process.

The role in this strange affair of Pakistan and its military was, and would probably remain, a mystery. Peter Savage, during his several tours of Afghanistan, had little doubt that the Taliban received supplies and shelter from Pakistan, and the border was never really closed to Taliban fighters. With Bin Laden's body now in their possession, the Americans had a difficult call to make - what to do with it. The last thing they wished to do was to upset the Islamic world by improper treatment of the corpse. The decision to commit his body to the deep within twenty-four hours of his death seems reasonable. As indeed was the accepted practice of washing the body and cladding it in a white robe before placing it in a weighted cover and sliding it into the sea.

The next day Savage went to his office at MI6. Already security at the building had been stepped up and Madge, his secretary, told him in hushed tones that she had been physically searched. The general feeling among the staff was that some retaliatory act was sure to follow somewhere in the world, with America and Great Britain as the most likely places for an act of revenge. Savage had a phone call from Sally Gunn to say she had been recalled temporarily to take part in raids by Special Branch on some likely suspects.

Peter Savage's Section Head came into the office and, using his old Naval charm, persuaded Madge to make some tea. Ralph Maitland sat back and accepted a chocolate biscuit from Madge, expressing his thanks by a gentle squeeze of her arm; tea quickly followed. "Well, how did you find Malaya and the island of Penang?" he asked, "And did Betty AG remember Carol Andrews-Dixon?" Savage nodded, adding: "The Academy is doing very well, but why the question about Andrews-Dixon? Mrs Ascott-Gallop thought she could be running a high-class brothel."

The old Naval Captain laughed. "Oh, she is, and it's very useful to us as well. These foreign diplomats get very loose-mouthed when they get a hard-on."

"But," asked Savage, "What use would some of Betty AG's girls be in that environment? They are not tarts." "Well," said Maitland, "Carol needs attractive young women as maids, and Betty's girls are intelligent and speak several languages. This could provide them with accommodation and some wages, while they pursued a more proper career in London."

Chapter Twelve

In several villages bordering Pakistan, Taliban supporters were clamouring for more information about the fate of a man they considered worthy of a place at the side of the prophet Mohammed. If indeed the Americans had murdered him, no doubt he would now be in Paradise with the blessing of Allah. The intermediate Taliban leaders were urging caution on any plans to seek revenge on their enemies. They should be patient; at this moment in time the Americans, British and their NATO allies would be on an extremely high state of alert, expecting an instant response to the death of a much-loved and respected former member of Al Qaeda and leader of the Taliban.

Peter Savage was not surprised that by late afternoon on this very busy day he was told to go out to his old stamping-ground in Afghanistan and sniff out some leads from old contacts. Savage immediately asked for the services of Sally Gunn but was told that, for the time being, she was unavailable. He was reassured that an agent would be available in Afghanistan who was up to the task and known to their posse of informers. At seven o'clock that evening Savage was at the RAF Afghanistan staging-post in the west of England.

With a stopover in Cyprus of two hours it was almost midday when they arrived at the main airport just outside Kabul. Peter Savage had just collected his suitcase when he was conscious of someone standing just behind him. Her perfume immediately betrayed her sex and Savage turned round to be greeted by a young and very attractive Royal Military Police Captain. She smiled and spoke: "Major Savage, I am Jill Downey and I am to assist you on your mission to this bloody awful country." Savage took the proffered hand; it was soft, but the grip was firm. "Thank you, Captain," said Savage "It's been a while since I was last operational here and you are clearly up to speed."

"Oh please call me Jill; it's more convenient as we are going to work together."

"In that case, call me Peter," said a smiling Savage, beginning to appreciate how attractive his new helper was. Jill Downey collected a Land Rover from the vehicle pound and within ten minutes they were driving through the gates of an army base.

"We have rooms in the large house over there; it was formerly the Governor's summer residence back in the days of the Raj."

The young Captain showed Savage to his room, explaining that they were all en-suite courtesy of the Americans, who had occupied it up to a year ago.

"No doubt you have worked with the Yanks before; they won't rough it unless they have to," added the smiling young officer.

"And where is your room, Jill?" asked Savage quietly.

"Oh, we females are on the next floor with a metal locked gate as a passion protector, and only we females have a key."

Savage unpacked his case, had a shower and changed into well-worn lightweight khaki drill shirt and trousers. He went downstairs to the Mess and helped himself to a cup of tea; then he was joined by Jill Downey, who had been talking to a rather large and very overweight Army Major.

"The boyfriend?" enquired Savage.

"Good God no," said Jill "The bastard thinks that as a Major he can order me to drop my knickers on command."

"Oh," said Savage "One of those."

At that moment the Major came over to their table. "Well," he said "We have a stranger in the Mess; please identify yourself."

Savage slowly got to his feet before saying: "My name is Peter Savage; who are you?"

"I am Major George Ferris-Patterson and I am just completing my second tour in this damn awful place." Savage smiled. "You have some catching up to do, my friend; I completed three tours in Helmand and not behind a bloody desk, either; but time is on your side."

Ferris-Patterson's face went almost purple and he abruptly turned and stormed off out of the Mess.

"I've made a friend," said Savage. Jill Downey burst out laughing. "My God, Peter, that was priceless; but, be careful - the man is a total shit."

The "total shit" was at that very moment already on the phone to the Base Commander, enquiring if the person from London was entitled to use the Officers' Mess, or would he be more at home in the Senior NCOs' quarters? Colonel Ralph Horsfield gave an exasperated sigh. Clearly, Ferris-Patterson had come off second-best in the two men's initial encounter. "Now listen to me, George. Major Savage is a Royal Marine Officer at present serving with MI6; don't obstruct him, give him all the assistance you can. He is here to deal with some very dangerous people. Is that clear?" A somewhat deflated Major Ferris-Patterson put down the phone after agreeing to give their guest his full cooperation. He needed to be patient; his chance would come.

The next morning Peter Savage and Jill Downey, in an unmarked Land Rover and both casually dressed, made their way to the daily market on the southern outskirts of Kabul. Both, however, were armed, with each carrying a Browning automatic pistol and two spare magazines. Both weapons were carried in waist holsters, deliberately on show to deter any opportunist thief. Savage was looking for a certain face operating a cold drinks bar from the back of a very old Citroen van. The van was there, but a very young lad was in charge. Savage waited until no one was nearby and asked: "Is your father here today?" "He is over by the livestock market trying to sell our two goats," said the boy, a smile spreading over his cheeky face. He then leaned forward and whispered something in Savage's ear. Savage responded by wagging a disapproving finger in the grinning lad's face.

Jill Downey went to ask what was being said, but Savage shook his head and they moved over to the livestock area, where Savage quickly spotted his target. "Make a fuss of those two scruffy goats, Jill, as if they were supreme champions." Jill did as she was asked. By now, Savage and the man had already made eye contact. Jill was playing her role to perfection; she grabbed Savage's arm and said: "Oh buy them for me, darling; they will keep the garden clear of weeds." The owner quickly quoted an outrageous price and a small crowd gathered to see if the foolish foreigner would be parted from his money in purchasing these two flea-riddled champions. The deal was quickly done and Savage handed over a fistful of the local currency to cheers from the small, amused crowd.

Jill, Savage and the goats' previous owner made their way back to the Land Rover, where the man tied the goats securely up in the back. A whispered conversation took place between the man and Peter Savage, the substance being "Osama Bin Laden's death was still being investigated by those now in charge; they had ordered no response until death was confirmed and the time was considered right for revenge". On the drive back to the British base Savage told Jill what had been said. "I thought that would have been their response, Peter." Savage agreed, and then asked: "What are you going to do with these bloody goats of yours?" Jill ignored the question, and then enquired what the young boy had said about her in the market. Savage smiled but said nothing, and she asked again. Savage sighed. "He said he could get a good price for you among the elders of his village and that you have a wonderful arse." Jill looked quite shocked and Savage added: "In fact I quite agree with him."

When they reached the compound the gardeners were given the goats to use in an attempt to keep the grass down. They were also told to treat the animals kindly and to wash them free of fleas. They looked at Savage in bewilderment, saying: "All goats have fleas, Sir." Savage gave a patient smile. "I know that, but please do as the Madam wishes." The two men bowed their agreement and gently led the goats away to begin a life undreamed of with their previous owner, who undoubtedly had more fleas than they had.

Over the next few days Peter Savage took Jill with him when he visited the local police compound, where he was permitted to look through their library of photographs of suspected terrorists. He took copies of several of the suspects and their details then told Jill these men were no longer in Afghanistan but frequent visitors to a mosque in Central London. Later that day a note was handed in at the Guard Room, addressed to the "Lady and Gentleman who bought two goats from me at the market". Inside, the writer asked if he could see them again as he had two female goats which would make their two males' lives happier.

The next morning they returned to the market and found it much busier than on the previous visit. Savage explained that this day the wives had accompanied their husbands, probably because the men were going to a meeting that had been called to take place after midday prayers. They quickly found their contact, accompanied by his cheeky-faced son and his heavily-pregnant wife. A whispered conversation took place over the condition of a bucket of heavily-bruised tomatoes; no goats were mentioned. Savage was told an instruction had been issued that no action in response to Osama's death was to take place until an order was given. As payment, Savage paid over the odds for the tomatoes and the bucket. As they left, Jill was very conscious of the cheeky boy's eyes fixed firmly on her bottom and was now feeling just a little flattered by the attention.

When they arrived back at the base Jill asked what they were to do with the bruised tomatoes. Savage smiled. "Oh, they are for your goats; if they get through that lot it should loosen them up a bit." While Savage made his way over to the wireless room to send his intelligence to London Jill took the tomatoes over to the gardeners, who failed to understand why the goats should have them. And when Jill had left they tipped half of them into their mutton stew, commenting that: "the English woman must be mad but she has a nice arse".

Later that evening a signal for Peter Savage was received from London. It suggested he should go to the Pakistan border after a visit to a friendly face in Kandahar who had a list of names for him. The next morning Savage and Jill left for Kandahar in an Army Air Corps helicopter. A curious Major George Ferris-Patterson watched them depart, grinding his teeth at the lack of progress he was making with Jill Downey and certainly not wishing them the best of good fortune on this journey.

About this time the United States began to show the world some of the tapes they had found in Bin Laden's hideout. It was, of course, a propaganda exercise designed to show the previous leader of Al Qaeda in the worst possible light. In response, as the Americans had not shown Osama's body being buried at sea, the followers countered with their own version of his disposal. They stated Osama's body had not been washed or clad in a white garment as their religion demands. He had been placed in the weighted canvas bag still blood-stained and in his badly-soiled clothes, accompanied by a freshly-slaughtered pig and, as a final insult to the followers of Islam, the weighted body-bag had been placed in the sea less than ten miles off the coast of India in an area where deposits from a large city's sewer system was known to settle.

When Savage and Jill arrived in Kandahar he was unable to contact his friend and they decided to go straight to the Pakistan border. They were given accommodation in the local police barracks, where conditions could be best described as 'basic'. They spent a frustrating three days there; since the capture and killing of the Al Qaeda leader there was reluctance to offer any cooperation to their British and American allies. On the afternoon of the fourth day they returned to Kandahar and stayed in some decent quarters in the Military compound. Savage's room was next to that of Jill Downey; he decided to see just how friendly she would be after dinner that evening.

Chapter Thirteen

In Washington the elation over the disposal of Osama Bin Laden had been tempered by the various reports from the Special Forces that had been corrected, not just once but now several times. The President was privately furious that the first account, then the second and even the third had needed to be corrected in front of the World's press. The Special Forces team that had carried out this operation could not be blamed for the confusion. The blame lay with the Defence Department being a trifle over-ambitious in filling in the gaps following the Special Forces' brief initial report. The Department, though reluctantly accepting the responsibility, claimed it acted in the best interests of providing the American people with as much information on the dramatic event as possible.

There was now another bridge to repair and that was with the confused and embarrassed Pakistani Government. Already the International News Companies were questioning how it was that the Al

Qaeda leader could live under the noses of the Pakistan Military for so long without them knowing. Some reporters were even saying that if the Pakistanis knew, then so did the American Military. The Americans wanted this story dead; too many questions remained unanswered and needed to remain so. Some severe flooding in parts of America received more publicity than would normally have been the case, and the death of a well-loved and respected Spanish golf legend added the final touches. With Osama Bin Laden and the confusion of his death now consigned to the inside pages, pulses at the White House resumed their normal but far from regular beat.

One or two experienced commentators suggested that the date for American Forces to withdraw from Afghanistan could now be brought forward. The suggestion was of course made in such a way that, without saying so, it demanded a response from the Pentagon, and it got one. It was announced that four units on standby to go to Afghanistan would now be placed on the reserve movement list. With money now an ever-constant topic, the savings this would make were considerable.

With no publicity at all, a top-ranking American diplomat was sent to Pakistan to pour some oil on the troubled waters. His wife, as always, accompanied him; her beauty was legendary and so was her sexual appetite. Up to eighteen months ago he had been Ambassador to Pakistan, having served there for two tours of duty. When he had completed his last tour there was a good deal of speculation on who was the greater loss. It was common knowledge at the time that the Ambassador's wife was having an 'affaire' with at least two top Government officials and one of her husband's chauffeurs. Within ten days of the couple returning to Pakistan the diplomatic harmony was almost totally restored. The diplomat was slightly greyer and thinner of hair, but his wife looked magnificent.

Back in Kandahar any hopes Peter Savage had of becoming closer to his attractive colleague had been frustrated by the good lady's natural cycle. They were then both kept busy when they were shown some of the tapes that had been recovered by the Americans at Osama's villa. The tapes showed an ageing Bin Laden, long past his best as an orator. They also suggested he had reluctantly handed over control to his deputy. This information would never be released to the world by the Americans as it made the capture and death of Osama less of a triumph than the Americans wished it to be seen as.

In Kabul all Allied personnel had been warned to be on their guard. Al Qaeda and the Taliban were eager to make a gesture of defiance to the world at large. Two officers, one British and one American, had been tasked for the last few months with giving talks on their country to local schoolchildren. Security for the pair was not considered a problem, as the schools were all within the Kabul City boundary. The British officer was a Lieutenant in the Grenadier Guards; his regiment was not serving in Afghanistan at the time. Mark Diamond was the son of a prominent Jewish banker; his two brothers were in the family firm but Mark was not considered 'banking material'. He fancied a military career so he had been given a Commission in the Guards entirely due to his father's influence.

The American officer had an entirely different background, his father being a guard on the New York underground system. His mother worked at a bakery and he had one sister training to be a teacher. Harry Lovelock had studied hard to pass the entrance exam at a Military School that was certainly not a threat to West Point. Despite their different backgrounds the two men got on well and, though not close friends, enjoyed working together. This particular day they were to give a two-period talk at one school then, after lunch, go to a girls'-only school and repeat the morning exercise.

In accordance with the latest rules both men were armed, but instructed to carry these weapons discreetly on the school visits. A ten o'clock they arrived at the boys' school and parked their Land Rover under a large tree. They then made their way into the building, which seemed very quiet for a busy boys' school. Finding no one about, Harry Lovelock turned to his colleague and said: "Well, the young buggers appear to have the day off." As the two young officers made their way back to the entrance three men stood by the main door. Both young officers went to remove their pistols from their briefcases, but it was too late.

Three automatic weapons were pointed at them and they were ordered onto their knees and their weapons were removed from the cases. Their wrists were tied together crudely by wire, as were their ankles. Rough gags were forced into their mouths, making it difficult to breathe; then they were dragged into the back of a three-ton lorry. They were quickly covered with bags of animal feed and then, with two of the men in the back and the third driving, the lorry lurched out of the school yard. Ten minutes later they had cleared the outskirts of Kabul; they had passed through two check points but were not detained for inspection at either.

At just after four in the afternoon they reached a small village on the edge of some large hills. They were then dragged out of the lorry and down some rough steps under a whitewashed house. The gags were removed for just a few minutes while they were given some greenish but welcome water to drink, then the gags were put back in place again. The cellar door was closed and they were alone. Both tried to speak but the gags were too tight, so they just lay there, already fearing the worst. By six in the evening it was almost completely dark and the two men heard voices outside the cellar door. They were not only the voices of men, but also those of women. The two men shuddered with fear - they had heard how cruel Afghan women could be.

Two donkeys dragging a small cart were to be their mode of transport; blindfolded as they were, they did not know this. Both men were seized by two or three men and thrown into the cart; their appearance raised to a new level the excited chatter of the women. It took almost two hours for the cart to complete its journey. Their accompanying group of men and women was now silent, adding an air of menace to the situation. Their gags and blindfolds were removed and the tight wire binding to their ankles was removed. The moon was now providing some light and the two men realised they were at the junction of three dirt tracks overlooked by two very old and gnarled olive trees.

One of the men in authority spoke and the men drew back from the two prisoners and the women moved forward, their hands outstretched. The young Guards officer was selected first for their attention and knives suddenly appeared in three of the women's hands. Trembling now with fear, Mark Diamond waited for the thrust of the first knife. It came, but not as expected. His clothing was being cut from his body and, at the first sight of bare flesh, there was a sharp intake of breath from the busy women. Within a few minutes his body was completely naked; even his boots and socks were removed. His body was prodded and one woman carefully examined his private parts, her remarks causing the watching women to laugh and add comments of their own.

Harry Lovelock was next. He struggled in an attempt to protect his modesty, but his clothing was still cut from his body and his struggles had caused some cuts to appear on his body. In normal circumstances he would have welcomed women seeing his naked form because he was well-blessed in the reproductive organ parts department. Several of the women grabbed his penis, making a variety of comments that provoked laughter from both men and women. Suddenly the leader of the group shouted something and the mood changed. Mark Diamond was seized and a rope was tied to his ankles, one end being thrown over a branch of the tree about twelve feet from the ground. Three men took hold of the rope's end and Diamond was secured to the tree, dangling head-first some three feet from the ground.

Harry Lovelock suffered the same fate, hanging from another branch, both men now some twelve feet apart. The women now began to collect small pieces of wood from around the wooded area and, with the men's clothing to provide added fuel, their intention became clear. Soon sufficient fuel was gathered to make a satisfactory and long-lasting fire under each man. To prevent the men creating movement by the swinging of their arms, their wrists were secured to their waists; then the fires were lit.

Mark Diamond was the first to react as his longish hair began to smoulder, and his screams became so loud a gag was thrust into his mouth. Two minutes later Harry Lovelock received the same treatment, reducing the response by both men to a sobbing moan. As dawn began to lighten the eastern sky the fires, having disposed of their fuel, began to cool. The two prisoners were examined. Both had badly burnt heads and blistered shoulders, but some life was still there. A comment was made and four women ran forward and began to hack off the men's private parts, penis first and then their testicles. Both men still managed to produce agonised, muffled screams and blood ran down their bodies onto their charred faces.

It was now time for the onlookers to depart and return to their village before any searching troops arrived in the area. The two-hour journey took thirty minutes less on its return, both donkeys anticipating breakfast as a reward for their night's work. Back at the scene of the night's torture Mark Diamond was the first to find a blessed end to his life. Harry Lovelock was just ten minutes behind him and, within minutes of their passing, the birds began to examine a rare breakfast and two local foxes had to be satisfied with the discarded flesh that had been removed by the women's knives. Later that day a small patrol of three armoured cars arrived at the scene and eventually the bodies were recovered. Later, three of the soldiers in the patrol had to be treated for severe shock.

Though badly disfigured, the two men were not difficult to identify and a decision was made on the circumstances of their deaths. Two senior American and British officers issued a joint statement that the two officers had died when their Land Rover had left the road during a dust storm and had landed upside down in a gully. Both men were trapped and, with their vehicle bursting into flames, they were burned to death. In an official ceremony their bodies were sealed in their coffins and flown to their homes, where they were buried with full military honours. Several reporters attached to the American Headquarters were anxious to find out the true circumstances of the men's deaths. They were invited to a private dinner with the Commanding General and, whatever their reward was to be, the official version of the men's deaths remained unchallenged.

Chapter Fourteen

Peter Savage went down to breakfast the following morning to be handed a signal ordering his return to London. Jill Downey joined him in devouring a generous plate each of sausage and eggs and the news of his recall. What annoyed Savage more than anything was the face of a certain Major displaying his appreciation at his rival's departure. As Savage left the Mess he paused at the table of Major George Ferris-Patterson. "I shall be back here within a month and if Jill tells me you have been a bloody nuisance I shall take great pleasure in rearranging your face. Is that clear enough for you?" Ferris-Patterson said nothing, inwardly raging that the threat to his person had been overheard by the occupants of at least two nearby tables.

The flight home followed the usual path via Cyprus. It was here, however, that Savage was handed a message changing his route home. He was instructed to go to Rome and meet a member of Italian Intelligence who wished to share a recent discovery with him. Savage recalled meeting the man some years; ago his name was Carlo Pontifini. The Italian agent possessed film-star good looks and had a string of stunning girlfriends who tolerated one another without any sign of jealousy. It was early evening when Savage arrived in Rome and he was met by the British Embassy's head of security. Arrangements had been made for him to spend the night at the Embassy and meet Carlo at ten o'clock the next morning at the Central Police Station.

Savage had stayed at the British Embassy before; it had a reputation as being a cold place lacking goodwill, and it had not changed. The present Ambassador and his wife were aloof from their staff and a class system involving accommodation, food and leisure was applied to all residents. Savage was given a small bedroom with a shared bathroom, but no waitress service in the large dining-room. He realised he had been graded one step from the bottom, where he would have had to share a bedroom with a driver or security guard.

At thirty minutes before his meeting, Savage arrived at the police station and was shown into a waiting room full of solicitors anxious to see their clients. Carlo was ten minutes late and his usual charming self, and he greeted Savage like an old friend. Once in Carlo's office and with his attractive secretary pouring some coffee that smelt wonderful, they got down to business. "We are having trouble with our new technology, my friend, and I suspect your country is as well." Savage thought for a moment then said: "Is it just computers or mobile phones, or does it include military equipment?" Carlo smiled. "All of those things, my friend; I have asked the

French if they are having trouble and they say not, but you know what liars the French are."

Savage told the Italian that the Americans were having trouble with their Drone aircraft in Afghanistan and as they were using them in Libya NATO was holding its breath. Carlo nodded. "Yes, my friend; if those Drones target a hospital instead of a military bunker the shit will really hit the fan." Peter Savage suggested to Carlo that the Italian Defence Ministry should talk to the British MOD. Carlo Pontifini gave a rueful smile. "That's just what I suggested, Peter, but our people think the British would have to ask the Americans first and their answer would be no. Perhaps you could use your influence in this matter?"

The two men discussed other matters and then Carlo said: "I am going to take you to lunch, my friend; it is owned by the father of one of my girlfriends." The small and very smart restaurant was within sight of the Trevi Fountain; it was full, but a table was quickly found for Carlo and his guest. The owner came over to greet them. He was smiling and said: "My future son-in-law is always welcome." Savage thought he detected a threat there somewhere, but the owner was beautifully mannered. Both men decided on the fish. It was a good choice. A bottle of white wine was disposed of, then some ice cream and coffee. No bill was presented and Carlo on leaving spoke to the owner as his "Papa to be".

Savage was dropped off at the British Embassy and quickly made arrangements to fly to London that evening. Nobody suggested he should prolong his stay. The flight to London was uneventful and, in welcoming heavy rain, Savage took a taxi to his apartment in the security complex, looking forward to seeing Sally Gunn again. Savage was in his office by 8-45 but found his secretary, Madge, already there chatting away to Sally Gunn like an old friend. He spent the next thirty minutes telling them both of his time away; no mention was made of Jill Downey. Leaving Madge to sort out the in-tray of mail, Savage and Sally Gunn went into the back office to exchange what information they had.

Later that morning the Minister of Defence was confiding in his Principal Private Secretary the substance of a meeting with the Prime Minister on the matter of defence costs. "Was the meeting minuted?" enquired his PPS. "No," replied the Minister "But the PM is a gentleman; he would not deny what was said." The PPS said nothing, wondering how someone so naive could reach the heights of Minister of Defence. His Minister then went on to say that the Cabinet, under the PM's guidance, was looking at defence costs and looking at cheaper alternatives to some of the highly expensive systems the military was demanding. With his Civil Service survival alarm sounding, the PPS enquired if the Minister would care to give an example of one of the PM's ideas.

"Well, the Typhoon's weapon selection system costs the earth. We could install the cheaper American model as used in their updated F17 (S) and halve the cost of equipping a squadron," said the Minister. "And what does the Royal Air Force say?" enquired the PPS quietly. "Oh, the usual rubbish that we would be devaluing the Typhoon's superiority over the F17 (S). Typical of the boys in blue; at times their brains fly higher than their aircraft." After suggesting to his Minister that he should record word for word what had been said at the meeting with the Prime Minister, the PPS hurried away to consult with a trusted Treasury colleague. By noon the PPS had discovered from his Treasury friend that any arrangement for the purchase of additional equipment for the Eurofighter was subject to contract. If one country decided to install a different system in its aircraft below the original contract price, a penalty clause in the contract would be invoked making the alternative more expensive.

By late afternoon the PPS had placed all this information on paper for his Minister; he also had an arrangement with MI6 to keep them informed of any likely Government difficulty over international agreements regarding defence. At MI6 Peter Savage was informed of the suggested change of an equipment supplier because of the spate of equipment failures which he was investigating. When he mentioned to the Captain how fortunate it was having a PPS willing to pass on sensitive information, the old Naval Captain laughed. "We did the man a favour when he was in bad odour with his department; he will repay us for the rest of his career. That is the price of the quality service we provide."

The next morning before Cabinet, the Defence Minister managed a brief meeting with his Prime Minister; it did not go well. In a frosty exchange, the Prime Minister expressed his disappointment that his choice of Defence Minister was not as fully acquainted with his brief as he should have been. The Defence Minister's day was not helped when he received a confidential briefing from the Royal Air Force that in operations over Libya the Tornados were proving more effective in adverse weather conditions than the highly-expensive Typhoon. The final point was on the decreasing number of Tornados being available for operations because, as they were being withdrawn from service, no replacement parts were now available to keep them flying.

Another matter that was drawn to the attention of MI6 was the agreement between the Treasury and the Defence Ministry when the Libyan operation was first launched. The Treasury, under pressure from the Prime Minister, agreed to underwrite the cost of the operation, much to the relief of the Defence Secretary. However, in the small print this excluded the increased replacement cost of all the bombs, missiles and weaponry used. As an example: one British submarine had fired, to date, twenty Tomahawk cruise missiles. Each missile cost one million pounds, but the new replacement cost was twenty-five per cent higher.

As is usual in these matters, each branch of the armed forces suggests to the Minister that it is better value for money. Another ploy widely used at this time is that the Army and the Royal Air Force claim that, since they have two Royal Princes in their ranks, any further cuts should be borne by the Navy so as not to put the Royal Princes at risk. Peter Savage and Sally Gunn paid a visit to the Procurement department in the Ministry of Defence. They were not exactly welcomed with open arms. They were kept waiting for fifteen minutes in a shabby room containing a table piled high with very old magazines, and not a chair in sight. At last they were shown to a

Head of Department, who made several phone calls to establish their identity and the purpose of their visit.

Eventually he was satisfied and enquired: "How can I help?" "Who is responsible for checking the quality of the product you buy and who does the testing?" asked Peter Savage. "Sometimes the Military insists on a demonstration before any contract is signed, but our people are widely experienced in this area and have years of experience," replied the Department Head. "Are they all completely at ease and familiar with modern technology?" enquired Sally Gunn. The Department Head ignored her question, so she repeated it. Reluctantly, he replied that they were. Peter Savage, sensing a weakness in this particular area, enquired as to the average age of the team which was responsible for this new technology.

The Department Head shuffled some papers around before replying, and then grudgingly admitted that the team members were all in their early sixties and close to retirement. Savage sensed blood and suggested that it was time for a younger team to be in place, more familiar with the great strides that computerised equipment had taken. The Department Head very reluctantly agreed to give the matter some thought. Savage and Sally Gunn left the building, once more frustrated with the lack of urgency in the Civil Service to modernise itself and move confidently forward into the new century.

As the pair made their way back to the MI6 building, Savage enquired if Sally would like to take in a film that evening, followed by some supper. After a brief hesitation she agreed, asking what film he had in mind.

"Your choice," said Savage. "Just one thing; I am not too keen on Westerns." At seven o'clock they met outside a three-cinema complex. Sally was dressed in a smart, light-blue dress short enough to display her excellent legs. Savage smiled a greeting and asked what film she had chosen. She pointed to a poster; it advertised the latest in the Pirates of the Caribbean series. Savage struggled to hide his disappointment; after all, what fearsome pirate would wear eye shadow? He had to admit, though, he did enjoy the film and an added bonus was the contact his knee made with Sally's leg during a frantic sword-fight.

The supper went well until a couple who were two tables away had a violent disagreement with blows being exchanged, the female being the more aggressive. Sally Gunn talked some sense into the pair, with the support of Savage's large frame as extra security. The manager showed his appreciation by giving them both double brandies with their coffee. Afterwards, Savage saw Sally to her taxi and was rewarded by a kiss on the cheek. As the taxi drove away, he was suddenly glad he had not tried to go further. They had a good working relationship and a closer intimacy could spoil it. At four in the morning his phone rang; it was the Captain. "You are on a nine-thirty flight to Pakistan from Heathrow. Sally Gunn will be with you; we need to rebuild some trust with the Pakistan military and they seem to like you."

Chapter Fifteen

Some ten miles east of the border with Afghanistan is a Pakistan military base situated just a few miles from the Pakistan town of Khan. In the base a convoy of twenty fuel tankers was taking on its precious loads of petrol. If all went well they would the next morning make a daylight and heavily-escorted drive through the Gomal Pass into Afghanistan. Over the last few months the Taliban had destroyed over a hundred of these tankers, essential life-blood to the British and American vehicles on their vital operations. The Gomal Pass is midway between the more famous Khyber Pass and the lesser-known Bolan Pass. When the tankers reached the Afghanistan border, six British armoured vehicles would escort them on their journey.

As an added precaution, a small team of SAS troopers had been inserted by helicopter on one of the highest ridges in the pass, giving in daylight an early warning of any Taliban activity. With their agents and supporters everywhere, the Taliban knew of the convoy's departure time and also of the presence of the SAS spotters. The unexpected plays a part in every activity known to man, including the SAS. During the night one of the three-man team had become very ill with a suspected appendicitis and at first light the entire team was evacuated by helicopter, watched by the curious Taliban.

The now absence of their SAS spotters should have been passed to the Officer Commanding the escorting Mastiff 2 armoured vehicles but for some reason this was not done. At a few minutes past nine o'clock, the twenty-tanker convoy reached the head of the Gomal Pass in bright sunshine. Within minutes, the Pakistan military escort handed over to the British responsibility for the tankers and the journey into Afghanistan began. The six armoured vehicles were strategically placed in the convoy, two at the front, two in the middle and two at the rear. The first five miles were covered without any sign of Taliban activity. Captain Ryan Diley of The Blues and Royals, in command of the armoured vehicles and the convoy, slightly unclenched his tense buttocks; another fifteen miles and they would be clear of the bloody Gomal Pass.

In common with most passes through mountain ranges, the width of the road or track varies. In places, vehicles can pass one another with room to spare, but in other parts the track allows just one vehicle through with just inches of clearance from the sheer rock sides. Disaster struck as the leading tanker was edging its way through one of these narrow passages. Captain Diley was standing in the open turret of his armoured vehicle watching the tanker ease its way through, when an explosive charge was detonated. It immediately caused the tanker to explode with stunning force, filling the narrow passage with a fierce fire of burning fuel. Tanker two tried to reverse away from tanker one, but the third tanker was slow to reverse and tanker two became another victim of the conflagration.

A shocked Diley ordered his signalman to inform the other armoured vehicles what had occurred and to open out the remaining convoy's spacing and assume a defensive posture. Meanwhile, he could do nothing until the fire died down and he could tow the wreckage clear. With some reluctance he also sent a signal to Headquarters informing them of the situation; his Colonel would be furious and he was an unforgiving man. At the midway point in the convoy, Sergeant Rex Foster was busy persuading the drivers of the so-far undamaged tankers to return to their vehicles, with the assisted threat of his cocked automatic rifle.

Thirty minutes later the smouldering wreckage of tanker one was towed a hundred yards forward to a wider part of the pass. Tanker two followed. Its driver had scrambled clear and was now a passenger in tanker three. The driver of tanker one could still be seen behind the wheel of his vehicle, just a charred skeleton but still at his post. The convoy resumed its journey, desperately hoping the trauma was at an end; and, for twenty minutes, it was. Captain Diley, still standing in the turret of his armoured vehicle, was scanning the route ahead with his field-glasses. Something caught his eye and he focused on the spot just in time to record a rocket-propelled grenade heading his way.

He desperately tried to scream a warning, but it was too late and he and his two fellow crewmen died instantly. Two more missiles came their way, one going through the cab of tanker three the other striking tanker four. Both tankers exploded and the following tankers reversed, desperate to avoid incineration. The Taliban now played their trump card and their fighters concealed back up the pass launched more grenade attacks. Tankers nineteen and twenty were soon in flames. The remaining tankers were now trapped, with burning vehicles blocking each end of the Gomal Pass. Over the next thirty minutes, eight more of the tankers were engulfed in the spreading flames, their drivers unable to move their vehicles forward or back. With survival in mind, the drivers left their vehicles and were heading rapidly back towards the Pakistan border as fast as their feet would carry them.

Sgt Rex Foster was uncertain what was happening at the front of the convoy, until the crew of the other Mastiff 2, which had been caught in the exploding tankers' fire and destroyed, informed him of the fate of Captain Diley. Of the twenty tankers which had started the journey only six now remained, but were unable to move in either direction. Foster realised he was now the senior man and therefore responsible for what was to happen next. He first contacted the commanders of the two Mastiffs at the tail of the convoy. They told him that as soon as the wreckage cooled they would begin to clear a path for the remaining tankers to return to Pakistan with the four surviving Mastiffs.

The next step was to inform his headquarters of their situation. Like his late Captain Diley, Sgt Foster knew what a difficult man the Colonel could be. He passed the dramatic situation report and was told to "Wait On". A minute passed and the Colonel came on the net. The voice was brusque. "Why can't you continue into Afghanistan? We desperately need this fuel, Sergeant." Foster explained that there was more wreckage to move in front of them than to the rear, and most of the remaining tanker drivers had absconded in the direction of Pakistan. "You should have shot a couple of them and the rest would have stayed," said his unsympathetic Colonel. He then added: "You are to stay where you are and protect the six remaining tankers. Tomorrow at dawn we will helicopter in some infantry. Is this clearly understood?"

Rex Foster acknowledged the order and passed on the Colonel's instructions to the other three Mastiff Commanders who, like him, were unimpressed by the decision. In the two hours of daylight remaining they managed to herd the six surviving tankers together and pull clear some of the obstructing wreckage. A night defensive position was established, with the six tankers in the centre of the four Mastiffs at a wider part of the pass. They had five tanker drivers with them; the sixth tanker would be driven by a Mastiff driver who, previous to his army career, had worked for Pickford's International Removals.

All of the Mastiff 2 armoured vehicles were equipped with night vision equipment, which gave the soldiers a clear view of any Taliban trying to sneak close to their position under cover of darkness. As it was, the long night passed without further incident, the Taliban fighters perhaps satisfied with the number of tankers they had destroyed. The lightening sky was a welcome sight for all concerned and they celebrated the new day with a mug of steaming tea and some dry energy bars, the five remaining drivers sharing in the feast.

Shortly before eight o'clock the helicopters were heard and soon they were overhead, choosing a spot to discharge their cargo of eager infantry. Due to the narrow pass first one helicopter landed and discharged its troops then took off, allowing the second one to do the same. Their task now completed, the helicopters flew back to their base about thirty minutes' flying time away. Sgt Rex Foster reported to the Lieutenant in charge of the infantry, giving him chapter and verse on the events of yesterday's disaster. The Lieutenant expressed his sympathy and together the two men worked out a plan, first to clear the wreckage allowing the remaining tankers to resume their journey.

By eleven o'clock this was completed and, with the infantry scouting ahead of the convoy, the final leg of the journey began. It took all of three hours to finally clear the Gomal Pass and move into open country, with only the occasional threat from high ground to concern them. With the light beginning to fade they reached the main camp, where the tanker drivers wasted little time in transferring their precious loads to the underground storage tanks. Sgt Foster reported to his Squadron Sergeant Major, who at least gave him a smile of welcome before saying: "The Colonel is over at Battle HQ with the Brigadier; he will want to see you in the morning, so get some rest while you can." Rex Foster made his way over to the tent he shared with five others and spent the next hour writing out his report while the events were still fresh in his mind. He then showered and had his evening meal. By ten o'clock he was in his bed, waiting for sleep to claim him and put some comforting distance between him and the events of the last two days.

Chapter Sixteen

The Boeing 747 touched down gently in the afternoon heat of Karachi Airport. Peter Savage did his customary flexing exercises to rid himself of the flight stiffness, unaware of his companion's amusement. At last the plane came to a halt at the air bridge and there was the usual mad rush by the passengers to remove their belongings from the overhead lockers. Determined to avoid the rush in leaving the plane, Sally Gunn and Peter Savage were almost the last to reach the exit door, giving a nod of thanks to the cabin staff. As they passed through the passport check a hand was waved to them in greeting; it was Jali Mohammed.

Savage had to smile as Sally Gunn was given the full Mohammed welcome of fluttering lashes and soft brown eyes and the gentle kiss of the hand. Savage had to be satisfied with a firm grip of the hand and a brief "Good to see you again, Peter." They were taken straight to a waiting police Land Rover with the assurance that their baggage was being taken care of. They went first to a combined Police and Army compound where they were each given a cold, refreshing drink. Jali then gave them a summary of incidents involving communication failures and misbehaving computers, all attributed to the introduction of a virus of sorts.

Out of hearing of his army colleagues, Jali said the Pakistani Air Force had been obliged to ground some of its new fighters because of confusion in their Head-Up display when seeking target selection. Even Savage could appreciate the fighter pilots' problem when encountering a hostile threat, where a split-second reaction is the key to survival and combat success. After spending two hours at the compound, and with their heads beginning to spin, Sally and Peter were taken to their hotel to recover from their journey. Jali said he would pick them up at ten o'clock the next morning, his eyes seldom leaving Sally Gunn's face or figure.

Their rooms were on the fourth floor, adjoining and facing east. Their luggage was there, courtesy of a police driver. Both wanted a shower, a change of clothes and, most important of all, a brief rest before dinner. Ninety minutes later they went down together for their evening meal. They had been given a corner table; obviously Jali Mohammed had influence. The food was excellent, as was the service; then, when Sally Gunn looked around her, she commented that they were the youngest couple in the room. Savage laughed. "I began to wonder why it was so quiet. Now we can see why this lot will be in bed by ten-thirty."

After a good night's sleep they met in the dining room at eight-thirty. They had the place to themselves and the attention of the complete dining-room staff. Sally settled for fruit juice, toast and coffee; Savage decided on a cheese omelette and a pot of strong tea. After breakfast they took a walk in the hotel gardens then went to the hotel foyer to wait for Jali Mohammed. At exactly ten o'clock, Jali arrived in his own car. Sally got straight into the back seat and a clearly disappointed Jali had to be satisfied with Peter Savage sitting alongside him. Jali told them that they were going to see a new squad of police recruits pass for duty, then he would take them to the police laboratory to see some confiscated computer parts from China.

They watched with interest the passing-out parade. The young policemen gave an impressive display of drill following an inspection by the Minister of the Interior. Everyone then went into the main building for refreshments. It was packed with the proud families of the young policemen. Savage, who was talking to Jali, was suddenly aware his feet had left the ground and he was moving in a rush of hot air along the floor. His senses then were overcome momentarily by a huge concussion of sound. His previous military experience told him he was involved in a large explosion and he struggled to control and marshal his senses and start to look for Sally Gunn.

Savage managed to struggle to his feet, aware he was bleeding from cuts to the face and hands from flying glass. His still-dulled hearing could make out screaming and shouting for help. He realised he had been thrown about twenty yards from his original position when the explosion had occurred. He moved as quickly as he could, calling to Sally but barely able to hear his own voice. He then spotted her light-blue blouse and white linen skirt; she was half-covered by a fallen stack of chairs. He pulled the chairs to one side, revealing Sally lying motionless and pale. The angle of her left arm suggested it was broken. He called out to her twice, then again; this time an eyelid fluttered and she opened her eyes and said: "My God, Peter, look at your face; it's a mess."

Using the two sleeves of his shirt, Savage made a sling for Sally's damaged arm and then quickly checked her for other injuries. Apart from some minor cuts and scratches there appeared to be nothing more serious than her broken arm. He slowly and gently helped her to her feet then took her outside, where help was beginning to arrive, and sat her down in the shade. "I am going to see if I can find Jali; he was just in front of us and can't be far away," said Savage. Sally nodded and squeezed his hand. "I won't be long," said Savage, and moved back inside. Three stretchers were carried past him and a quick glance revealed they carried what remained of some young policemen.

A few yards further on Savage came across the body of Jali Mohammed. It was slumped against a wall, clearly hurled there by the blast. The position of his head to that of his body suggested his neck was broken; otherwise there was hardly a mark on him, but he was clearly dead. Savage spent a few minutes by his late friend. One consolation was that, whatever followed this life, Jali would now be reunited with his beloved wife who had died in childbirth. The ground floor of the building was now swarming with firemen and medics, so Savage made his way back outside to find Sally Gunn.

Savage could not see her anywhere, but a young policeman doing his best to control the arriving and departing ambulances said she had been taken to the nearby hospital. It was less than two hundred yards away and Savage slowly made his way there. Despite the number of injured being rushed through its doors, the hospital was coping very well. Savage approached a uniformed security guard who took one look at him and, ignoring his question about the whereabouts of Sally Gunn, summoned a nurse to take him for treatment. He was led to a cubicle where a motherly-looking twenty-stone nurse gently treated his injuries. She deftly removed small glass splinters from his face and scalp, sutured three deeper cuts to his neck then treated his hands and arms. As she finished she said: "No need to see Doctor today; go home and rest. Come back tomorrow if you need to."

Savage thanked her and went back into the lobby and enquired about Sally. He was directed to the second floor, where he found her in the passage outside the x-ray department. He stayed with her until she had been x-rayed and her arm placed in a plaster cast. Savage informed the doctor where they were staying and he allowed them to leave, with the instruction to get some bed-rest. Savage went outside and managed to find a taxi, and ten minutes later they were back at their hotel. The reaction of the hotel staff was astonishing, and they were almost carried to their rooms with such care as to suggest they were in their nineties. The manager came to see them both, assuring them that the hotel doctor would see them when he returned from helping at the Police College explosion.

Both Savage and Sally assured him that would not be necessary, as they had been given treatment at the hospital and just wanted some rest. They managed a couple of hours of sleep each, then washed and changed before carefully making their way down to dinner. As they entered the dining room several of the other guests stood up and expressed their concern. Both now had visible bruises to accompany their cuts and scratches; Sally stood out with her plastered arm. The waiters were most attentive- one insisted in cutting up Sally's food for her and even offered to feed her, but she quickly declined his kind offer as she saw the expression on Savage's face.

Savage put a call through to Madge in their London office, asking her to pass on the situation to their superior and saying that they would be taking a couple of days to recover. A very concerned Madge told them to be careful, stressing that clearly Pakistan was a very dangerous place. The next morning at breakfast, both admitted to feeling stiff and sore and Savage decided he would spend a leisurely day at the hotel. Sally, of course, had no choice. At ten-thirty there was a phone call for Peter Savage; it was from the Pakistan Security Headquarters saying that an agent would call. His name was Sami Sadique. Savage had met him before, about a year ago. Sadique was only about five feet tall but was almost as broad as he was tall and immensely strong with it.

Sami arrived about eleven. Sally had gone back to her room and, after shaking hands, Savage ordered coffee for them both and found a quiet spot in the corner of the main lounge. Sami Sadique told him Jali's body had been flown north to Peshawar for burial where his remaining family lived and his wife was buried. He also told him that the previous day's bombing had been committed by a suicide bomber and was thought to be in revenge for the death of Osama Bin Laden. The bomber had been disguised as a tea-trolley attendant and the curtained lower part of the trolley had contained at least a hundred pounds of explosive.

Savage knew his Department was still trying to come to terms with the fact that Osama had been living quite openly in Pakistan, and enquired how much Pakistan Intelligence knew of his hiding place. Sami smiled and said: "Oh, we knew where he was and so did the Americans; he was a sick man and no longer the Taliban leader, just a figurehead." Savage raised an eyebrow and asked: "Why the big, dramatic assault on his compound, then?" Sami Sadique shrugged. "It was all political. The American President appreciated something to divert concern about the American economy and the death of Osama Bin Laden did just that." "And what did Pakistan get out of it for its silence?" enquired Savage. Sami put his finger to his lips before saying: "A squadron of F17 fighters and some spare parts were all part of the deal."

For the next hour the two men discussed a problem which both countries were experiencing, and that was cyberspace attacks on their defence computer systems. In the last few days, America had warned that any country attempting cyber attacks on the American Defence Missile Shield would be considered to have declared war on their country and would suffer in consequence. "Where do you think the computer viruses are originating?" enquired Savage. "I think China could be behind most of it, but the production of these devices is being done everywhere; no country is immune," replied Sami.

Sami Sadique had never said a truer word. Not more than two hundred yards from where the two men were sipping their coffee, a former small, boot-and-shoe factory was under new management. Its staff had also changed from the aging men and women whose strong and battered fingers had fashioned boots and shoes at a reasonable cost to the local population. Now a young workforce of nimble-fingered school-leavers was busy making certain critical adjustments to a mass of brand-new mobile phones arriving from China.

Chapter Seventeen

Just before Sami Sadique was about to leave, there was a phone call for Peter Savage from London. It was Madge with instructions that Sally Gunn was to return to London immediately and recover from her injuries there. Savage was instructed to go to Afghanistan and contact the British Senior Intelligence Officer in the Kabul headquarters. Sami, who was still with Savage, offered to arrange a flight for Sally leaving the next morning. Savage agreed. Sami also suggested he had a small outing planned for the next morning and afterwards he would arrange for Savage to fly to Kabul.

The next morning after breakfast, Sally said her farewells and a police driver took her to the airport and saw her onto the plane. Sami took Peter Savage to the local military airfield and they boarded a police jet aircraft for their flight north to Rawalpindi. Savage was curious as to the mystery tour, but Sami smiled and would say nothing. One hour and forty minutes later they landed at Rawalpindi and transferred to a police helicopter for another thirty-minute flight. This time they landed at the main police headquarters in Abbottabad and Savage realised the purpose of the outing.

After a coffee, they were driven by Land Rover to the outskirts of Abbottabad, to a district known as Bilal Town. Everything now became clear to Peter Savage. Just a few weeks ago, on the second of May at one am, it was here that Osama Bin Laden was shot dead by American Special Forces. Their Land Rover stopped outside the walled compound familiar from the newspaper photographs; it was still under a heavy police guard. After a brief security check they were allowed inside. The place was a complete mess, with the ground-floor windows and doors reduced to a splintered mess. The indentations of bullets scarred the walls of the staircase to the first and then the top floor. Sami led the way into the room on the top floor where Osama met his end. It was so quiet, almost church-like; even Sami Sadique spoke in a hushed voice.

"That was where he died," said Sami, pointing to a large dark stain on the floor, "And that was where his wife lay after she was wounded in the leg." Savage again noticed many bullet indentations on the walls, some tightly grouped, indicating automatic fire. Most of the furniture was smashed, including a bed and torn paper and clothing littered the floor. Sami pointed to a shelf, saying that two weapons had been found there, one a semi-automatic rifle, the other a Browning pistol; no attempt had been made to use them.

"He could have been taken alive," said Savage. Sami nodded then said: "But nobody wanted that, including

Osama. Dead, his legend lives on and over the years his story will improve with the telling." With nothing more to be seen, the two men went back down the stairs and out of the building; both were silent. On the journey back to the police headquarters Sami told Peter Savage they would have a late lunch there and at three o'clock a helicopter would arrive to fly him to Kabul. Fish was on the menu, and very nice it was too. Later, the two men sat on the veranda drinking their coffee. At two minutes to three o'clock a British Lynx helicopter arrived. Savage and Sami walked out to it, Sami carrying his friend's case. The two friends shook hands and Savage climbed aboard and strapped himself in, donning the earphones provided.

Sami Sadique backed away and covered his face as the Lynx wound up to full power and climbed into the sky. A minute or so later the pilot informed Savage over his headphones that they were passing over the compound where Bin Laden met his end; Savage waved an acknowledgement. Forty minutes later they landed on the parade ground of the British Headquarters in Kabul. Savage thanked the pilot, grabbed his case and exited the helicopter, which soared away to the south. No one came to meet Savage and he walked over to the main Officers' Mess building, hoping to see Jill Downey.

At the Mess hallway desk a Corporal smiled a greeting. "You must be Major Savage, Sir, Captain Downey described you. I have given you your old room; here is the key, Sir. If you leave your case there I'll have it brought up to you shortly." Declining the offer, Savage took his case with him up to his room, wanting to freshen up before he met up with the delightful Jill

Downey. Thirty minutes later a refreshed Peter Savage returned to the ground-floor lounge and started to read a day-old English newspaper there. He was conscious of someone standing near him, so he put the paper down. It was Jill Downey. "My God, Peter, what have you done to your face?" she exclaimed "It's all cuts and bruises." Savage explained about the suicide bomb attack at the passing-out parade and that Sally Gunn had returned to England because of her broken arm.

Jill ordered some tea from one of the Mess orderlies and gave Savage the latest information about operations there, with particular reference to the fuel convoy fiasco. She also mentioned the American drone programme, where the targets had been selected but some faults in the weaponry used had caused them to go astray and strike the wrong buildings, killing innocent men, women and children. Peter Savage told Jill of the conversations he had had with intelligence staff in Pakistan and their conviction that computer programmes were being sabotaged by a virus that accompanied routine messages from mobile phones.

Jill wanted to shower and change before dinner so she went to her room. Savage decided he would rest on his bed for an hour - it had been a busy day. At seven-thirty Savage went down to the Mess, was given a table and waited for Jill to join him. He had passed the gross figure of George Ferris-Patterson but had not exchanged a word. Ferris-Patterson, however, was very curious when he saw the cuts and bruises on Peter Savage's face. As two Captains joined him at his table he pointed Savage out to them and said in a loud voice: "It would seem that our visitor has been caught in flagrante delicto by a furious husband." Jill Downey, who was passing the Major's table at the time and heard his remark, could not contain her anger. She paused and said: "For your information, Sir, Major Savage was at a Pakistan Police College when a suicide bomb attack took place and twenty people were killed. Your remarks, Sir, were not those of a gentleman."

As Jill took her seat next to Savage he noticed how angry she was and asked why. In a low voice, she told him. To her surprise Savage laughed. Further conversation was halted as one of the Mess waiters came over to take their order. Savage asked what choice they had, if any. "Oh yes, Sir," said the young waiter with a smile "It's a choice of pea soup or sardines on buttered toast for starters, main course is chicken and three veg, and a choice of sweet is ice-cream or apple pudding and custard." Savage looked across to Jill and said: "You choose; I can eat anything." Jill laughed and decided on sardines, main course and ice-cream.

After their meal, Jill took Savage into the Mess bar and introduced him to some of her friends, both male and female. The evening went well and Savage decided to say goodnight to Jill in the bar, conscious that Ferris-Patterson was watching them. "See you at breakfast," said Jill, adding "We have the day with the Intelligence Section." A storm struck Kabul and the surrounding areas during the night, giving some welcome heavy rain. At breakfast Jill Downey was already into her corn flakes when Savage appeared. He looked at the corn flakes without interest and helped himself to porridge from the self-service counter. After sausage, bacon and egg and two cups of coffee, they were both ready to start the day and wandered over to the car park, both wearing side-arms.

The Intelligence Section was a mile from the Mess and shared the site with the main armoury. A Captain and a Sergeant greeted them and took them to a room where, on a table, was a selection of items in common military use. They had one thing in common: all had been compromised by additional electronic aids. Peter Savage was handed a Global Positioning Monitor in common use by officers and NCOs. The device told them instantly where they were by providing a six-figure map reference of their present position. This was essential for casualty evacuation, resupply and avoiding friendly fire. "Take a reading, Sir," said the Sergeant. Savage held the monitor clear of his body at face level then checked the reading.

Savage was not surprised to find the reading corresponded with a figure on the wall which said: "You are here". "Now do it again, Sir," suggested the Sergeant, who moved over to the door with something in his hand. Savage took another reading; this time it was different. The Sergeant explained that the second reading had been affected by a mobile phone number and was at least a mile out, putting your men in danger of air and artillery strikes by friendly forces. "What part of the kit has been 'got at'?" said Savage "Is it the mobile phone or the GPS monitor?" "In this case it's the mobile phone," said the Sergeant "But some mobiles and I-Pads can compromise computerised systems such as the automatic pilot in aircraft without being changed from the manufacturer's design."

Savage did not say so, but he was well aware of this following the investigation into the aircraft that had discharged all its fuel over the Atlantic. The officer then informed them of the vulnerability of the new drone aircraft to conflicting commands from its controller and Taliban interference from near the target area. It was a morning well spent and most of the information fitted neatly with that gathered in Pakistan and elsewhere. After thanking their hosts, Jill Downey and Peter Savage returned to the Mess for lunch. Jill said she had something arranged for the afternoon and most of the next day, but had a couple of details to finalise.

She returned after about ten minutes, smiling. "We are going to join a patrol of three Mastiff armoured vehicles having a good look in the foothills near the Pakistani border." Savage, who was hoping for something more romantic, hid his disappointment well and hurried off to change into something more appropriate. They were to travel in the lead vehicle, commanded by a young Lieutenant whose father owned half of London. Each Mastiff had a crew of three, but four additional infantrymen had been added to Mastiffs two and three just in case they hit trouble.

Lieutenant Jamie Pinkerton greeted Savage politely, having been informed of Savage's background by Jill Downey. Having settled in the severely basic comfort of the Mastiff the Lieutenant reminded his guests that he was in command and they must obey his orders instantly. Both Jill and Savage nodded their agreement, Jill's eyes never leaving Savage's face, seeking his reaction.

Chapter Eighteen

Some twenty miles east of Kabul in the foothills leading to some majestic peaks, a group of twenty Taliban fighters rested outside a series of small caves that for a few days had been home. As part of their deployment they would spend another week in Afghanistan, causing as much trouble for the occupying infidels as they could. Then, unless they were told otherwise, they would cross back into Pakistan to spend time with their families before wandering forth again. Their leader was Azim Zaman, a fearsome figure who tolerated no weakness in his men. Azim had one weakness: he lusted after European women. This could be traced back to a school he attended as a boy. It was run by a charity and staffed by volunteers. His teacher had been Mary Goodchild, a widow from the north of England in her early forties, but still attractive and full-breasted.

Azim was a well-built fourteen-year-old and was in the last months of his schooling. One afternoon as the class dispersed, Mary Goodchild asked him to stay behind and help to rearrange the classroom. All went well until they moved a broken desk into a darkened store room. The door suddenly swung shut and Mary Goodchild attempted to squeeze past Azim to open the door. The contact between the two gave Azim the opportunity he had prayed for and triggered a surge of lust he could not contain. His hands moved everywhere, above and under his teacher's dress. Her shrill protests became feeble and, as her hand encountered his erect state, with a groan she breathlessly submitted to his demands.

Full intimacy may not have occurred as Azim was not completely sure how to proceed but, urgently guided by his now-willing accomplice, a full and satisfying coupling took place. Azim left school a few months later, but their affair continued for another three years until Mary had to return to England to nurse her elderly mother. Azim came under the influence of a radical Imam at his local mosque and he was guided to work in a local forge where, not only did they shoe horses and repair carts, but they made crude firearms. Over the next couple of years Azim joined the Taliban and rapidly progressed through their command structure. Several times he was involved in taking European hostages in Afghanistan for ransom. If they were female they were forced to submit to his demands and then to that of his colleagues.

From his present position, Azim Zaman had an unrestricted view of the rough terrain which led to Kabul, although the city was not in view. His chief interest was in a cutting that had been blasted by the Russians through a rocky ridge, shortening the distance to the Pakistani border. The cutting was about eighty yards long and was used by the military on its many patrols. Over the last two nights Azim and his men had mined the sides of the cutting, hoping to catch a supply convoy coming from or returning to Pakistan. Though Azim would prefer to be nearby when the explosive was detonated it was not necessary, as two pressure-plates, laid in the road and a couple of inches under the surface, would trigger the mayhem for him.

Some movement caught Azim's eye; it was at the edge of the heat haze which shielded Kabul and he raised his field-glasses to see what it was. It took little time for him to confirm what it was. What had caught his eye was a dust cloud raised by several vehicles some distance away. Azim turned to the men nearest to him and told them what he had seen. An excited babble of conversation was exchanged, partly silenced by Azim giving those closest to him a reproachful look. Ten minutes later Azim raised his field-glasses again; this time he had a clearer view. He was slightly disappointed; there were three armoured vehicles, probably Mastiffs. Unless he could disable at least two of them he would not engage them.

In the cramped troop compartment of the lead Mastiff it was hot and the air was filled with a fine, powdery dust. Jill Downey sat opposite Peter Savage, her knees raised slightly to adjust to the lightly-padded seat. They had been on the road for just over an hour; it seemed longer and Savage was beginning to wonder what Jill Downey was hoping to achieve in this expedition. A few minutes later the feet and then the rest of Jamie Pinkerton appeared from the box turret of the Mastiff.

"We will stop in about five minutes for a pee and a smoke." He smiled at them both. "Not too uncomfortable, I trust?"

Shortly after his visit, they stopped. The Mastiffs were about fifteen yards apart. About a mile ahead they could see the cutting they would have to go through. Pinkerton spoke to Jill Downey: "If you want a pee, Captain, please go over to the left; we will go to our right." Up in the foothills Azim raised his powerful glasses to his eyes and examined the armoured vehicles and their crews with interest. He counted eighteen, then nineteen figures. One figure in particular attracted a second glance; it was a woman. Azim watched with bated breath as the slim figure moved away from the main group of men and went round behind some large boulders. The reason became clear as she undid the waist-band of her trousers and lowered them and her undergarment. Azim had a clear view of her naked, white bottom as she squatted to relieve herself. He felt his heartbeat begin to race.

Jane took just a minute or so to empty her bladder but, in doing so, she had the most uncomfortable feeling she was being observed. As she adjusted her clothing she realised the watcher, if there was one, could not possibly be one of the soldiers from the patrol as they were all grouped between the Mastiffs. The driver of Mastiff Two was in earnest conversation with Lieutenant Pinkerton and, judging by the expression on the Lieutenant's face, what the driver had to say was not appreciated. Jamie Pinkerton came over to where Savage and Jill Downey had found some shade. "Got a small problem with one of the Mastiffs; its power steering is playing up. We could be here for an hour or all bloody night."

Savage looked at Jill Downey and shrugged. "Well, these things happen, but at least we have plenty of protection." An hour passed, then two, and the light was now beginning to fade. Savage walked over to where Jamie Pinkerton was watching the progress, or rather the lack of it, on the Mastiff's steering problem. The Lieutenant was clearly angry and it was beginning to show. He turned to Savage and said: "It's not the fucking power steering; it's a buckled coupling on the front pair of wheels." Peter Savage shrugged. "These things happen, Jamie; I take it the repairs will have to wait for daylight? We can't illuminate the scene or the Taliban will pick us off at their leisure."

Five minutes later it was pitch dark and the Mastiffs and their passengers were deployed in an all-round defensive position. The Mastiff's two machine-guns had one weakness: they could not engage a target within twenty yards because of the limit to which they could be depressed. Peter Savage and Jill Downey sat next to each other, leaning against a wheel on the inside of the Mastiff triangle. Everyone had enjoyed a mug of tea and some savoury energy biscuits before they took up their night positions. The night was cloudy, denying them the help from an almost-full moon. Lieutenant Pinkerton had been in touch with Headquarters by radio, informing them of their position but not requesting any assistance. It was all a matter of pride and standing on one's feet.

Back up the hillside the Taliban were on the move. This situation suited them and the threat from the armoured vehicles' weaponry would be restricted by darkness. Two of the Taliban's rocket-launchers were fitted with a night-vision sight; these had been captured from a police armoury in Kabul several weeks previously. It took Azim and his men ninety minutes to reach a position some two hundred yards from the cluster of Mastiff armoured vehicles. Azim sent eight men to the right of their objective; now both groups could fire without being a threat to each other. One of the men carrying a rocket-launcher moved a few yards further forward; he now had a clear view through his night sight of one of the Mastiff vehicles.

Jill Downey was almost asleep, leaning partly against Peter Savage with her head resting on her small pack. Suddenly a blinding flash and a huge explosion shook the ground, and one of the Mastiffs erupted in flames. Savage dragged Jill Downey under the vehicle they had been leaning against; both had drawn their pistols to defend themselves. Despite not having any clear target to aim at, the gunners in the remaining two Mastiffs sprayed several bursts of their 7.62 mm machine-guns in the general direction of where the missile had come from. To the frustration of Azim and his men the second rocket-launcher with a night sight misfired. The Taliban fighter, brave as he was, did not know the procedure for dealing with a misfire and kept pressing the trigger with, of course, a nil result. His problem could have been solved by a simple check of the rocket's wiring and a tightening of the terminals.

Over the next two hours the soldiers, with their night-sight equipment, had little difficulty keeping the Taliban at bay. In the explosion that destroyed the one

Mastiff, two soldiers had been killed and three wounded, one seriously. Azim realised his chance of a major success had gone and was already moving his men away in twos and threes. His mind could still picture the female's white bottom. What a trophy that would have been, but perhaps she had been killed when the Mastiff was hit. An hour later, although the soldiers did not know it, the threat to them had gone back into the hills. Jill Downey was still lying beneath one of the Mastiffs, the solid and reassuring figure of Peter Savage close by. In fact he was very close by, his lower half very firmly up against her bottom. She felt safe and, at the same time, thought a change of position was called for.

Over the last hour the cloud had begun to thin and it was with some relief that they were suddenly bathed in moonlight. Their attackers had gone and they all breathed a sigh of relief; night engagements are never popular, as they give the poorly-trained terrorist almost parity with a professional soldier. Savage had enjoyed the feel of Jill's well-shaped bottom firmly placed against his groin, and in different circumstances would have gone further. At last the first signs of a new day appeared, and in less than twenty minutes they were in full daylight and it was time for breakfast.

An hour went by and the movements of the day were decided. A recovery vehicle under escort would arrive shortly to tow the remains of the badly-damaged Mastiff back to base. Fortunately it was the vehicle with the buckled coupling, so they had two fully-operational Mastiffs to carry them back as an additional escort. Lieutenant Pinkerton asked Savage if he would join him and four of their protection team in inspecting the cutting ahead of them. He, like Savage, thought there was a connection between the Taliban and the cutting. With two of the soldiers keeping an eye out for any hostile movement, the remainder carefully edged their way along the road. Within minutes one of the soldiers found one of the detonating plates and, a few minutes later, the other was discovered.

It took another hour to locate the explosive buried in the cutting's sides, now completely free of its detonators, and another hour to remove it. The recovery vehicle and team winched the badly damaged Mastiff onto the flatbed truck. Room was found for the bodies of the two dead soldiers and, in a separate vehicle, the three wounded were receiving expert attention. It was just after midday when they finally reached the main camp. Jill Downey and Peter Savage both thanked Jamie Pinkerton for the adventure, and then went to their rooms to wash and change before having lunch.

That evening there was a 'sods' opera' in the Mess, where all present are expected to contribute in some way. When it came to Peter Savage's turn, he recited the crude version of the Felicia Hemans classic poem Casabianca, more widely-known As The Boy Stood on the Burning Deck. The Chaplain stayed only for the first stanza, which went: "The boy stood on the burning deck his affection not in doubt, The Captain close behind him with his willy hanging out." The drink flowed and the evening did not get any better in the quality of entertainment, but everyone found it relaxing, which was the purpose of the evening.

As Savage was about to suggest to Jill that they did a tour of the garden, the Mess Corporal handed him a message; it was from London. He was required in Cyprus and a plane was awaiting him at the nearby military airfield. He had to be satisfied with a sisterly kiss on Jill's cheek, and then he collected his kit and was driven away at speed.

Chapter Nineteen

The plane touched down rather heavily in complete darkness; it was just after four in the morning. He managed to cadge a lift to the nearby Military Headquarters, where a surprise awaited him - nobody knew anything about him. He managed a couple of hours' sleep in the lounge of the local Officers Mess, and then was first in the dining room for breakfast. The Mess Secretary cast a suspicious eye on Savage and demanded to know who he was, in the Mess unshaven, and having breakfast as well. All was well on production of his identity card with the MI6 logo, and Savage could not have wished for better treatment.

With a good breakfast inside him and the opportunity to shower and shave, the day had dramatically improved. A visit to the Headquarters Intelligence Section revealed he was just in transit. His destination later that afternoon was Benghazi, where he was to meet a French contact. Savage was advised to go to the military airfield and wait for his plane, as no one seemed quite sure who had been detailed to take him. At two o'clock a French military aircraft landed and he was quickly hustled out to join it. As soon as he was seated, the plane took off and headed west; at least he was going in the right direction.

After about ten minutes a rather large man dropped into the seat next to Savage's. He introduced himself as Pierre Sarte and he was a member of French Intelligence. Clearly, Pierre's lunch had included garlic and rather a lot of it; Savage was not averse to the bulb, but in reasonable amounts. "We have been given what you call the 'shitty stick', my friend," said Sarte in heavily accentuated English. "Our diplomatic friends in both governments, having launched the Libyan crusade to eject old Gaddafi, are now not too sure who leads the rebels and we have been tasked to find out." An air attendant handed them a glass of beer each and Sarte smiled his thanks and drained his glass in one. Savage sipped his beer before enquiring: "Have we been given a list of who our governments would like to be in charge?" "No," said Sarte "Only a suggestion of who they don't want, such as Al Qaeda, Hezbollah, Hamas and anyone from Israel."

The plane landed at Benghazi Airport after its ninety-minute flight and was immediately surrounded by a variety of four-wheel-drive vehicles. Their drivers and assorted passengers immediately demanded if they had brought any arms and ammunition. When they were told "No" they waved their arms in disgust and drove off, firing their weapons wildly into the air. Pierre Sarte and Peter Savage were greeted by a Junior French Diplomat, who took them in his car to a large and surprisingly clean hotel. They were given rooms on the same floor and advised to wear side-arms all the time, even in the dining room. After a coffee they were seen by two advisors, one British and the other French, who each had two Special Forces bodyguards from their own country constantly at their side.

Both advisors said the same thing: "We want you to get to know the rebel leaders, find out who gives them instructions, and then pass these names on to us." Later that evening, after a decent meal, the two men made their plans for the next day. They would drive down the coast road to Sirte, where there was some fighting, and start their search there. They were given two former Libyan policemen as interpreters and local advisors so as to make their task a little easier. As it was, all foreigners were treated with extreme suspicion, the only law here at the moment being the gun. At six o'clock they had breakfast, having slept quite well and, as the light improved, they started their journey in daylight; driving in the dark was a 'No-No'.

Every few miles along the road was a check-point of sorts, manned by a motley crew of badly-disciplined but heavily-armed members of the anti-Gaddafi freedom fighters. With the aid of their two former policemen each stop was brief but, when an enquiry was made as to "who is your leader?", it provoked a heated discussion and disagreement among the group. Several times they heard the sound of high-flying aircraft and then the very distant rumble of explosions; clearly, NATO was active day and night. They reached the eastern outskirts of Sirte at about two in the afternoon. The road was packed with rebel vehicles of all types. Some were mounting heavy-calibre machine-guns fitted in a most precarious fashion; others carried recoilless guns, equally a threat to the user.

Through their police interpreters they enquired where the battle headquarters were; they were met with confused stares. When the question was asked: "who decided when to attack Gaddafi's soldiers?" one man replied: "When there are enough of us, we all do." At this moment two mortar bombs exploded on the road ahead of them; chaos ensued with vehicles heading left and right into the desert and back up the road to Benghazi. They decided to stay roughly where they were, placing their vehicle on the eastern side of a badly-damaged cold store. There was the attendant smell of rotting vegetables and fruit, plus something else. A brief investigation provided the answer; just showing from under the pile of rotting vegetables were two pairs of feet. They did not investigate further.

With what breeze there was coming off the sea, they stayed where they were. They had a week's supply of rations, plenty of water, and their sleeping-bags to keep out the night desert chill. The four of them split up into two groups to eat and sleep. It was decided that each of them would do an hour apiece in succession to guard against intruders during the night. Pierre Sarte and Savage discussed the day and what they might do tomorrow; the picture so far was confusing, with apparently no one in charge.

When Peter Savage awoke he felt something was wrong. It was - there was no one on guard. He nudged Sarte with his knee; the Frenchman stirred, farted and went back to sleep. Savage checked his watch; it was twenty minutes to six, so he decided to allow his companions another twenty minutes' sleep. He took a badly-needed piss on the seaward side of the building and took a short walk to stretch his legs. A slight noise to the east attracted his attention; coming slowly along the road towards them was a convoy of vehicles. To be on the safe side, Savage woke his colleagues and suggested to one of the former policemen that he should greet his countrymen with a friendly word and a wave.

The man, known to Savage and Sarte as Ali, was not over-keen but reluctantly did as he was asked. The convoy returned the greeting but did not stop, moving a further half a mile before coming to a halt. After breakfast and a large mug of coffee each, Peter Savage suggested they move along the road and see what their freedom-fighter friends were up to. They joined the confused group and, after much shouting and the waving of arms, a decision was agreed. The plan was for a group of vehicles to drive into Sirte to see if they came under fire from Gaddafi's forces. Next to be decided was who the heroes would be. Savage saw an opportunity to gain some credibility and suggested to Pierre Sarte it should be them. Sarte gave a shrug of acceptance and told their interpreters to pass on the good news.

With what appeared to be good wishes ringing in their ears, they set off for the Sirte outskirts. Several of the outlying buildings were still smoking from the previous day's battle and gave the intrepid quartet some cover in their cautious advance. One of their interpreters urged them to stop by a large, badly-damaged building, saying: "This was police station." Leaving one policeman in the Land Rover they went inside; it was not a pretty sight. Bodies of fighters from both sides lay on the floor; obviously the building had changed hands more than once over the last few days. They decided it was time to call up the rebel forces and their driver drove back to do so. Ten minutes later a score of vehicles arrived and discharged their anxious warriors into Sirte. Firing wildly into the air the men advanced down the road which led to the town centre and on to the seafront, thinking Gaddafi's men had fled. They were mistaken.

About thirty men of the rebel advance party reached a road junction which, amazingly, still had a set of functioning traffic-lights. With some glee they shot the lights to pieces, gaining considerable satisfaction in the act. It was short-lived and several bursts of machine-gun fire from surrounding buildings left at least ten of their number motionless in the road. Another seven or so were helped away by their shocked comrades to some nearby shelter; clearly, Sirte was still defended by Gaddafi's men. Peter Savage and Pierre Sarte were not surprised at this outcome; the Libyan army was not that highly regarded but, man for man, they were more than a match for these rebels.

Pierre Sarte had a mobile phone and rang a number he had been given. A countryman quietly asked for a map reference. Sarte consulted his mobile phone, which had a Global Positioning

Facility. He gave a six-figure map reference, adding: "Two hundred yards due west of that." One of the interpreters was told to tell the rebels that an air strike was on its way, and everything went deathly quiet. Twenty minutes later there was a rush of sound and three ground-shaking explosions occurred to the left of the shattered traffic-lights. The rebels stared open-mouthed at the sky from where this horror had been launched, but there was nothing to be heard or seen.

By nightfall half of Sirte was now in rebel hands. Savage and Pierre Sarte were now treated with great respect as men who could do things. The four men managed to make some tea and feasted on some tinned meat and biscuit before it became completely dark. Two groundsheets, extended from the side of the Land Rover, provided cover from any night shower. The two men were now comfortable in each other's company and able to be frank over any plans, deciding to leave the next day open for discussion. A spectacular night thunderstorm deprived them all of at least two hours' sleep, but the rain cooled down the air and made the conditions more pleasant. In the morning during breakfast, Pierre's mobile rang. The message was brief and he turned to Savage, saying: "They want us back in Benghazi, my friend."

Chapter Twenty

The two men and their interpreters started their journey back to Benghazi just after nine o'clock the next morning. After about thirty minutes they were stopped at a road-block and asked to take a badly-wounded man into Benghazi hospital. He had suffered a chest wound and was having difficulty in breathing. As usual in these cases he was more comfortable sitting up than lying down and travelled to Benghazi propped up between Savage and one of the interpreters. He managed to explain the circumstances of his wounding in short gasps. A friend cleaning a rifle nearby had forgotten to unload his weapon and had accidently shot him at close range. Savage and Pierre Sarte exchanged glances; a poorly-trained man with a weapon and live ammunition deserved to be given a wide berth.

Their journey back to Benghazi was completed in good time, the wounded man being waved through at most check-points. At the main hospital they carried the gasping man into the emergency section, where he was placed in a wheelchair and given oxygen. They then carried on to their hotel, where their interpreters dropped them off. A rather officious junior Foreign Office chap came over to Savage and enquired who had given him authority to call in air strikes. Savage, anxious to keep his temper, said nothing and walked over to the tea bar and obtained a mug of strong tea. He was pursued by the irritant, who demanded an answer. Pierre Sarte joined Savage and asked what this little shit was after; Savage told him. Sarte turned to the official and said: "Go back to the nursery and play with your toys; we are engaged in men's work."

An hour later they were both summoned to see the Senior British Diplomat and his French colleague. They were both questioned closely about the rebel Command structure. Both men said the same thing: "There isn't any." The two senior diplomats looked at one another; the French Official asked, in perfect English: "But who decides when and where to attack Gaddafi's forces?" Peter Savage spread his hands in a gesture of puzzlement, and then said: "All these decisions are made on a collective basis, like a show of hands." The two officials looked at one another, the British member saying: "But our governments need to talk to the rebel leaders about the future of a free Libya once Gaddafi is removed from power."

After a lengthy discussion, the men agreed to meet after the evening meal. Savage, who was drinking another mug of excellent tea, looked at Pierre Sarte and said: "We will be off again in the morning, my friend; that I can guarantee." Later that evening Savage was proved right; their mission was put to them in such a way that it would have been churlish to refuse. The French Official, who was the senior of the two, stressed the importance of their mission and suggested that if they could bypass the small coastal towns and reach Misrata they may find some authority there. Pierre Sarte looked at Savage, who shrugged and then nodded his head. Both men knew Misrata was about one hundred and eighty miles from Sirte and, apart from the threat from Gaddafi's forces, they would be at risk from NATO air strikes.

With the two officials picking up the bar bill at ten-thirty that evening, Savage and Pierre agreed to try their luck as asked. It was decided they would leave with the two former policemen as drivers and interpreters after breakfast at eight-thirty. As an additional safety factor, they would carry an electronic beacon that would declare them as friendly to any roaming NATO aircraft. The next morning, after a full breakfast inside them, the four intrepid seekers of information left on their mission. They carried a week's supply of NATO All Forces rations and a generous amount of bottled water. The two Officials were not there to wish them "God speed", but a box containing the electronic beacon was.

At two in the afternoon they took the road to by-pass Sirte; rebel fighters at a check-point there warned them that, under cover of darkness, some Gaddafi supporters were bringing in supplies and reinforcements. One of their policeman interpreters suggested they drove about five miles into the desert and pick up a road built by the oil companies, which led to Misrata. It seemed the safer option and they took his advice. The one drawback of choosing this cross-desert route was stretches of soft desert sand, and it took almost to nightfall to reach the oil company road. They spent the night one hundred yards from the road in a hollow. Their bellies were full with a NATO steak-and-pasta meal, washed down in Peter Savage's case with a pint of strong English tea. The others chose coffee.

They each in turn stood a two-hour watch. The night was quiet, and their sleeping-bags kept out the desert chill. They breakfasted between six and seven, then took to the well-made tarmac road. They drove with caution between thirty and forty miles per hour, with Savage scanning the road ahead with his powerful field-glasses to give them an early warning of trouble. Just before noon, a column of black smoke in the distance suggested Misrata was still in dispute. An added complication was brought to their attention by one of the policemen, who explained that this oil road joined the coast road on the western side of the port, which was still in Gaddafi's hands. Deciding to take no chances, they left the oil road and headed north-east, which should bring them out on the safe side of the town.

As they drove carefully across the desert trying to avoid any obvious soft patches, a great shadow flashed over their heads accompanied by a great blast of air and sound. The Typhoon fighter aircraft, disappointed to have found a friend, screamed into a high, climbing turn to search for prey elsewhere. The Land Rover's four occupants gave the electronic beacon which was secured on the dashboard a grateful glance, realising how close to death they had been. Another thirty minutes and they came back onto the coast road, emerging into a virtual traffic jam of rebel vehicles. They were viewed with suspicion until one of their interpreters explained who they were. This was aided by one of the policemen's many relatives, who waved a greeting and gave a sweaty embrace to his cousin.

At this point a very smart man in combat fatigues appeared on the scene and gave a sequence of instructions to the hundred or so men leaning against the motley group of vehicles.

His instructions were carried out and Peter Savage and Pierre exchanged glances, realising this was the first evidence of authority in the rebel ranks. The man then turned his attention to Savage and Pierre Sarte; he addressed his enquiries to one of their interpreters, clearly curious to know who they were. He seemed satisfied with the explanation then turned to Savage and Sarte and said in very clear English: "Have you come to help or just to stick your fucking noses in to what is the Libyan people's business?"

Savage glanced at Pierre Sarte, who gave a slight nod of the head for Savage to continue. Peter Savage explained to the man that they were trying to find the leader or leaders of this anti-Gaddafi movement so their governments could offer help. The man laughed then said: "Up to three years ago I was a Major in the Libyan army, but I resigned to take up a senior security post with a major oil company." He paused and took a sip of water then went on: "When this trouble started I decided to try and organise some order in the rebel forces, and in particular the Misrata area; so far we have held our own, even when they used tanks."

At this point some mortar bombs exploded on the road some hundred yards away and there was a rush for cover in the desert sand. They lay together in the shelter of a rocky outcrop. The man said his name was Sol Narwaz; Savage asked if he was aware of who would replace Gaddafi should the rebellion succeed. Narwaz shook his head; he then said: "It could become tribal as it has been in the past, with a separate East and West Libya." Then he added: "If a group of so-called new leaders for Libya turn up suddenly in Benghazi at an opportune moment, the men who are doing all the fighting are not likely to embrace them with cries of gratitude and loyalty, in fact just the opposite."

They stayed where they were for another hour or so then Sol Narwaz offered to take them into Misrata. They left their Land Rover in the safe hands of their two interpreters and followed the former Libyan Army Major into the badly damaged town. "We hold the east, they hold the west," said Narwaz. "And the docks?" enquired Pierre Sarte. "Oh, that's no man's land," said Narwaz "In the early days we had an agreement to evacuate both sides wounded by ship, but it did not last; both sides broke the truce so it stopped." A stream of rebels was making its way into the town and an equal number was leaving. Narwaz noticed their interest and said: "Its hand-over time; each group does twelve hours in town keeping Gaddafi's men in check; then they take a break."

They stayed where they were while the change-over took place then with care followed Sol Narwaz back to their Land Rover. It was time to eat before the light failed, and they offered their new friend some of their NATO rations. Sol Narwaz returned to his headquarters after the meal with his new friends, and Savage and Pierre Sarte started to prepare for the night. It was about two in the morning when the Gaddafi forces launched their attack, a hail of mortar bombs falling on the rebel positions. Trained soldiers would have stood their ground, knowing that an attacking force is always at the greatest risk against a stubborn defence. But the rebels did not have that discipline and fled to the outskirts of the town, leaving behind their clothing and reserve ammunition.

Savage and Sarte joined Sol Narwaz on the road leading out of Misrata, watching as he tried to contain the rush to seek safety in the surrounding countryside. Narwaz worked hard and at last had some success with his reserves in forming a defensive line at the town's edge. Nobody managed any more sleep that night and all were grateful when the first faint light of dawn appeared in the eastern sky. Savage, Sarte and the two policemen started to prepare breakfast and boil water for tea and coffee. As the sun came up, shame-faced rebel fighters returned from the safety of the open desert, now eager to return to the positions they had deserted.

Sol Narwaz spoke to them all in the shelter of a badly damaged garage; he urged caution and uttered not one word of criticism. In small groups the rebels hurried back into town and within two hours were, in most cases, back in their previous positions. Clearly the soldiers of Gaddafi had not taken advantage of their enemy's retreat during the mortar attack. Peter Savage and Pierre Sarte had some talking to do, then a decision to make, in the light of the conversation they had enjoyed with Sol Narwaz. It was clear to them both that any ideas NATO had that the rebellion against Gaddafi was well-organised were false. It also left the door open for a group of opportunists to seek advantage in the developing chaos - the lure of oil and its international currency is a powerful incentive.

The two men talked for over an hour and decided to return to Benghazi and acquaint the two Foreign Office officials with the intelligence they had uncovered. They managed to find Sol Narwaz and informed him of their plans. He smiled and thanked them for their courtesy in confiding in him and, with a firm handshake, wished them their God's guidance and protection. He also suggested they return to Benghazi the way they had come, using what daylight remained in the day. Fifteen minutes later they were on their way across the desert to pick up the tarmac road and, with any luck, reach their old camp site before it got dark.

Chapter Twenty-One

The journey back to Benghazi took almost thirty-six hours. Twice they came under small-arms fire but did not return it, assuming it came from nervous rebel soldiers. The hotel was just waking when Peter Savage and Pierre Sarte were demanding their room keys from an elderly night porter. The two men treated themselves to a shave and a hot shower before going down for breakfast in laundry-fresh clothes. Pierre Sarte had eaten his two croissants well-coated with butter and honey before Savage was half-way through his bacon, egg, mushrooms and sausage. Drinking three cups of black coffee, Sarte gave Savage the chance to finish his breakfast at about the same time and wait for the appearance of their government's advisors.

Eventually the two men appeared and waved their two adventurers to join them at their breakfast table. As the story of their visit to Misrata was the same for both teams, Savage started it off and Pierre Sarte finished it. The British Delegate wiped some boiled-egg yolk from his lips before enquiring: "You say there is no evidence of any chain of command amongst the rebel forces?" "None

whatsoever," said Savage, with Pierre Sarte nodding his agreement. The two men then went on to say that in Misrata they had witnessed the only occasion in their venture into rebel territory where one person seemed to be able to control any of the rebel fighters. And the fact that he was a former Major in the Libyan Armed Forces said it all. In essence the rebel army was a bunch of enthusiastic amateurs, completely lacking in leadership and any command structure.

This news went down like a lead balloon with the two delegates, both of them aware this was not what their respective governments wanted to hear. More coffee was ordered and Savage and Sarte were interested to hear what the two Delegates were going to agree on in the joint message to their Government. The two men appreciated the company of their two adventurers and bounced a few ideas off them with the enquiry: "Would you think that was half true?" When both Savage and Sarte shook their heads, the two Delegates tried again. When Savage suggested that they told their political masters the truth, both the Delegates laughed. The British Delegate smiled at Savage and Sarte. "You two gentlemen are brave and determined former soldiers; you seek out your enemy and dispose of him. With us, we seek out the opposition and convince them to meet us half way; it is called diplomacy."

Later on that morning, a trio of new faces arrived and were treated with great respect by the hotel staff and were referred to as members of the newly-formed rebel government. Peter Savage and Pierre Sarte viewed the men with some suspicion, wondering where they had come from. Savage managed a word with the British Delegate and enquired who these men were. "We have been told to talk to them, though NATO has not given them full approval; but it is a case of half a cake being better than no cake at all."

It was late that afternoon when Pierre Sarte knocked on Savage's room door. He came in and sat on a chair by the window and delivered his message. The French and British Delegates wanted them to escort the three freshly-arrived gentlemen to Sirte. The gentlemen had expressed a wish to talk to the rebel commanders in the field. "What fucking commanders?" said Savage. Sarte raised a hand. "Be patient, my friend; let us see what happens when they realise the true situation in Sirte. It should be quite an eye-opener." It was agreed that at eight o'clock the next morning Savage and Pierre Sarte would escort the three newly-arrived members of the new rebel government to Sirte. They would travel in two vehicles, with Savage and Sarte's Land Rover leading the way.

The next morning, due to the three men being late risers, it was almost nine o'clock before they managed to get on the coast road. The rebel government members insisted on having the two former policemen as their driver and protector. Savage and Pierre Sarte were not aware, of course, of the stream of questions the two former policemen were subjected to on the journey as the government members demanded to know why more people were leaving the Sirte area than seeking to capture it. On arriving at the outskirts of the large town it seemed to Savage and Sarte that the rebels controlled less of the area than they did a week before. The next two hours were at first confusing and possessing some humour; the following hour developed into something far more dangerous.

With their frustration growing by the minute at their inability to find anyone remotely in charge, one of the provisional government members was openly critical of the rebels' fighting ability. There was a brief, stunned silence then one of the rebels with a bloodstained and bandaged arm thrust his automatic rifle painfully under the chin of the government member. A struggle ensued, with Pierre Sarte and Savage reluctantly going to the aid of the new government member, who was helped back to his vehicle with blood pouring from his nose. One of the three clearly possessed some sense and two bottles of medicinal brandy were handed over as an apology. Peace now restored, Peter Savage enquired: "What do the gentlemen wish to do now?"

At this point their minds were made up for them, as a salvo of mortar bombs landed less than a hundred yards away. They eventually stopped a mile back along the road and one of the former policemen lit up the gas stove and made some really excellent coffee. Even the bloody-nosed government member, who was still visibly shaken, regained some humour as they all sampled the coffee and a couple of sweet, high-energy biscuits. Two of the rebel fighters had received shrapnel wounds in the mortar bomb salvo and the new rebel government members offered to take the two back to Benghazi hospital. This gave them an excuse to leave the scene with some honour and, in fact, satisfied everyone. Pierre Sarte and Savage were particularly pleased to get these three creeps off their backs.

Though Savage was not aware of it until later, the generous act of the three rebel government members was cruelly rewarded when one of the wounded rebel fighters, still in shock, was violently sick over them during the return journey. It was almost six in the evening when Savage and Pierre Sarte walked back into their hotel. From the lounge a small, white hand waved a greeting. Savage gave the waver a second glance and smiled; the day was ending well, the hand belonged to Sally Gunn. Savage reluctantly introduced Pierre Sarte, who positively drooled over Sally's hand, kissing it twice as he muttered "Enchanté, Mademoiselle."

Pierre Sarte left to shower and look his best for dinner. Savage enquired how the arm was. Sally said: "Fine," and revealed an elastic bandage as support. She then passed on some information regarding the Libyan Operation. It seemed that the Government and the Ministry of Defence were at odds over the progress, or lack of it, towards the fall of Gaddafi. The MOD wished the news releases to be accurate and not overstate the progress made. The politicians were not satisfied with this and called on the military to express more optimism. A veteran General contacted the Prime Minister and bluntly stated: "If we are to lie about the progress of this operation I suggest all the daily situation reports had best come from Downing Street."

At dinner that night Sally Gunn attracted quite a lot of attention, most of it from Pierre Sarte. Peter Savage decided to let Sally handle the amorous Frenchman; after all, she had dealt with the sweaty-fingered colleagues in her last police squad. Dinner completed, the resident group of musicians started to play and Pierre whisked Sally onto the floor, showing what an excellent dancer he was. Savage sat sipping an after dinner brandy, waiting for the chance to show Sally he was no slouch on the dance floor either. The explosion, when it came, was from the food serving area; a blinding flash followed by a rush of hot air and a mind-numbing blast of sound.

Sally's reaction was an instant cry of "Oh, not again," then she fell over the large and now unconscious figure of her dance partner. She was aware Peter Savage was at her side and together they dragged their French companion through the door onto the outside terrace. "You OK, Sally?" asked Savage as they checked Pierre Sarte for any obvious injuries. Sally assured him she was fine. At this point Pierre Sarte sat up with a groan, rubbing a golf-ball-sized swelling on the right side of his forehead. He looked at Sally and almost smiled as he asked: "Did I go too far?" Savage burst out laughing.

Order was soon restored and, apart from a few cuts and bruises, no one was seriously hurt; then the cause of the explosion was revealed. It was not a terrorist bomb, but a gas bottle that had been placed too close to the hot-plate that it served. With the dining-room staff working flat out to restore their pride and joy, most of the hotel guests moved into the bar area. Savage obtained a drink for himself and Sally and suggested Pierre should get an early night to recover from his accident. Sarte nodded and gave Savage a look that lacked any brotherly love, kissed Sally's hand and walked slowly away to the lift. Sally sighed and said: "What a sweet man." Turning then to Savage, she said: "I think I'll hit the sack, boss; it's been a long day."

Savage finished his drink then the rest of Sally's, and was about to go to his room when one of the three rebel government officials came over and sat at his table. Without any preamble he said: "I need to speak to someone urgently in Misrata; would you take me there tomorrow?" Savage spent the next twenty minutes explaining that it would be a three- or four-day trip and that he must clear it with his Department Head first. "I don't want the Frenchman with us; just one of those former police guys for extra security," said the prospective rebel government official. It took Savage just a few minutes to sound out the Foreign Official on whether he should go or not. The Official nodded and said: "Yes, go; and find out all you can from any meeting."

At six o'clock the next morning the Land Rover, its police driver and four days' rations were waiting at the hotel entrance. Savage, standing by the vehicle's wing, glanced at his watch. At that moment the rebel government official hurried down the steps to join them. A brief nod of the head was his only greeting and the Land Rover, now with its full load aboard, moved off to the coast road with the first trace of daylight lifting the night's gloom. In the first hour they were trapped in a slow-moving convoy on its way to Sirte. But when the convoy stopped for a break they were able to accelerate past, ignoring the curious stares of the convoy's occupants. It was just after eleven o'clock when they reached the fork in the road to by-pass Sirte.

As on the previous visit they left the coast road and headed inland to pick up the tarmac road that led to Misrata. The police driver attracted Peter Savage's attention and pointed to the recognition beacon he had clipped to the dashboard. Seeing the rebel government official's puzzled look, Savage explained the purpose of the beacon and that it was essential for their safety. The official smiled and nodded his head as they drove deeper into the desert, seeking the oil-men's tarmac road.

Chapter Twenty-Two

Less than one hundred and fifty miles off the Libyan coast the French submarine Coutance was settling into the routine of HMS Triumph, which it had replaced. The Triumph was heading for Gibraltar for a two-week rest from operations and the chance to refuel, re-supply and re-stock its complement of missiles. Captain Jean Resaul of the Coutance was not exactly delighted with his assignment; his wife, Monique, had been born in Libya. Her mother was French but her handsome father was Libyan and a great supporter of Colonel Gaddafi, and both still lived in Tripoli. When Resaul had been informed of his assignment he had acquainted the base's Admiral with his family difficulty. The Admiral had listened, nodded his head in sympathy and let loose a great rumbling fart which echoed down the passage of the headquarters block. He then turned to Captain Resaul. "Do your duty, Captain; it is what your country requires. Your wife will understand; women always do."

Jean Resaul rested on his cabin bunk. At midnight they were due to fire two of their missiles at a suspected radar centre a mile from the Tripoli city centre. Resaul sucked on a sweet; he was only too well aware that the target was just half a mile from where his in-laws lived. In the submarine's missile room, three of the deadly weapons were being readied for the night's operation. It was custom and practice always to have a reserve. Senior Weapons Officer Jules Vernoux was checking each missile's guidance system. After the initial launch they would be guided by satellite, following each missile's set Global Positioning System. Over the Libyan coast, a high-altitude monitoring aircraft would ensure the accuracy of the strike. At least that was the theory. It was becoming increasingly obvious that what worked well in drills did not always work in operational conditions.

Resaul was beginning to drift off to sleep when a tap on his cabin door brought him quickly back to the present. In answer to his command of "Enter," his Executive Officer appeared in the tiny cabin's doorway. "My Captain, I am sorry to disturb you but we have a case of sexual assault in the Communications Section." Captain Resaul let out a groan; he had just two females on board and both worked in the Communications Section. He sighed and asked if it was the one with blonde hair and big tits. His Executive Officer shook his head. "Oh no, Sir; Petty Officer Le Mange was caught buggering Junior Telegraph Operator Paul Martell." "Oh, for fuck's sake," said the Captain "Was it consensual? If it was, we can fine them both and issue a warning." "It's more serious than that, Sir," said the

Executive Officer. "Le Mange forced the boy into the act with some violence."

Ten minutes later the matter had initially been dealt with. Le Mange was confined to his Mess on open arrest awaiting transfer to a shore establishment, and young Martell was in the sick bay. At nine in the evening the Coutance moved closer to the coast of Libya; the firing point was to be ten miles off the coast. At fifteen minutes to midnight the submarine surfaced; its position was checked against the satellite and the routine for firing its missiles commenced. At one minute to midnight the short firing sequence began and, on the button of midnight the first missile was launched, its flaming trail visible until it disappeared into the low cloud. Ten seconds passed and missile number two was fired. As at this point the submarine could be at risk, it now sped north but remained on the surface.

At thirty-five thousand feet over Tripoli the NATO aircraft which was packed with unbelievable technology tracked both missiles to the coast then inserted final target corrections. Confident, the young Weapons Officer watched as missile one neared its target, registering in red on the plotting board. To his surprise and concern it passed the target, swung back towards the coast and then dived to the ground and exploded. The controller switched his attention to missile number two - it was behaving at the moment as it should. He made a fine adjustment to its track directly onto its target then, as before, instead of diving onto its point of detonation it swung left for sixty seconds and dived to the ground.

The aircraft's Weapons Officer reported immediately to the Captain of the NATO aircraft. "Captain; Sir; both missiles have overrun their target and exploded." The Captain's voice showed the minimum of irritation as he enquired: "You had them under your control at all times, Mr Beven?" The Weapons Officer took a deep breath before replying "Yes, Sir, but both missiles ignored my signal to 'complete' and travelled another thirty seconds of forward flight." The Captain paused then asked: "Any idea what they could have hit?" There was no reply and the Captain repeated the question, this time with an edge of authority to his voice. The Weapons Officer eventually replied, very softly: "Sir, the area around the target is all residential."

Back at NATO Headquarters the failure of the mission had been flashed by the Captain of the aircraft responsible for any final adjustments to the missiles. The Duty Senior Officer let loose his feelings. "Oh, for fuck's sake; who checked the missiles over in the pre-firing check? It's fucking kids' stuff; what were the Brits up to?" He was quickly reminded that they were fired by the French submarine Coutance; the Brits were on rest and re-supply. The Duty Senior Officer gave the matter some thought then decided what action to take. "Instruct the Coutance to return to base. I want a thorough check made on the missiles still aboard and those in store in France." He then poured himself a cup of coffee and rang the Senior NATO officer on the base.

Lieutenant General Mark Lovell was fast asleep, snuggled up to the delightful backside of his new young wife, when the shrill ring of his phone brought him abruptly back to the land of the living. He picked up the phone and said abruptly: "Yup." He listened to the message once then said: "Run that past me again." It was, and he let out a deep sigh. "This is just the fuck-up that the anti-operation Libyan lobby have been praying for. Wake up the propaganda boys and get something from them to muddy the waters in the morning."

In Tripoli the convoy of ambulances was still queuing at all of its three hospitals to discharge the seriously- and lightly-injured survivors of the two explosions. Members of the international press had been collected from their hotels and taken to the hospitals to witness this latest atrocity committed by the NATO murderers. Adam Wright, a senior reporter with News International, turned to a yawning colleague. "Is this for real, or part of a Libyan attempt at 'black theatre'?" "Oh it's for real, my friend; the Gaddafi regime cannot believe their luck. Two submarine-launched missiles missed whatever they were aimed at and landed in a highly-populated civilian area."

Over the next four hours the reporters and television crews had ample opportunity to witness the victims of this latest tragedy. By morning an official list had been published of the dead and injured. Sixty-two dead were confirmed and ninety-six injured, most of them seriously. Even NATO headquarters did not query the figures. The propaganda experts came up with some ideas but these were quickly rejected; as one senior commander said: "This time we have just got to take it on the chin and express regret."

On the French submarine Coutance the Captain was again resting in his cabin as the submarine returned to its pre-firing position. There was a discreet tap on his tiny cabin's door. The Captain called "Enter" and one of the submarine's two females entered and handed the Captain a sealed envelope. "The Communications Officer stated 'Captain's eyes only', Sir," said the female rating. The Captain took the envelope from her, realising of course it had to be the one with blonde hair and big tits. The message was short and far from sweet. It stated: "Return to base at best speed; do not touch your remaining stock of missiles. They and your mission to be subject of inquiry."

Ten minutes later the Executive Officer and Navigating Officer had been called to the Captain's presence and informed of their new orders. "We are to return to Toulon at best speed and submerged. I calculate this will take about forty-eight hours; do you agree, Mister Navigator?" Rene Albert agreed with his Captain, suggesting they could make the return journey just a little quicker. "No; forty-eight hours will be fine," said the Captain. "Oh, and one thing more; I want a totally accurate list of the crew who had access to our missiles since they were embarked at Toulon, and an armed, double guard on the missile room until we reach Toulon."

With the submarine Coutance returning to its base, it left the NATO command without an underwater weapons platform. The Americans had several submarines in the Med, but their President's instruction to limit their involvement in the Libyan situation prevented their use. Britain and France were not willing to ask for American help in this area, as a rejection would be seized on by those countries already regretting their involvement in the Libyan operation.

In Tripoli the international telephone system had, since the start of the NATO operation, been severely restricted to its residents. But after the previous night's "slaughter of the innocents", as the national daily paper had reported it, the phone lines to Europe were once more open. One of the first persons in France to receive a phone call was Monique Resaul; it was from her parents. In the ten-minute, tear-filled exchange they told her of the loss of life less than a hundred yards from their home, and the deaths of their dear friends and their three young children. When the phone call had finished Monique rang one of her closest friends, whose husband was another submarine commander. It was then that the full impact of the tragedy was realised and Monique became aware that her beloved husband had come close to killing her parents. She went to her room and cried for an hour.

Twenty-four hours into its return journey, and yet another tragedy shook the crew of the Coutance. The forward shower room was reserved for the two female crew members from 0600hrs until 0630hrs each day. Ami Delgado entered the shower room and switched on the light; then she carefully removed her clothes and pulled back the shower curtain. Her horrified scream was heard by everyone in the forward part of the submarine. Swinging from the reinforced shower-head feeder-pipe was the naked, and very dead, body of Chief Jean Le Mange. First on the scene was her large-breasted friend, who made sure Ami's naked form was discreetly covered as she was led away from the horrific scene. The remains of Le Mange were placed in a canvas body-bag and put into the guarded missile-room on the Captain's instructions.

Submariners are a superstitious lot and Captain Jean Resaul realised that the events of the last twelve hours would require the submarine to be formally cleansed when it reached Toulon. This duty would fall on the two naval base Chaplains. It would be discreetly done, with just the crew to witness the event. The Captain, of course, had been unaware of his wife's telephone call from her parents in Tripoli. If he had been, his mood would have deepened even further, but he had that problem to confront when he reached home.

ChapterTwenty-Three

Peter Savage, his driver and the rebel government member had finally reached Misrata. It was almost midday and the scene did not appear to have changed much since his last visit. There were several groups of rebel fighters sitting in some of the damaged shops' doorways. Savage asked their police driver to ask one of them to take a message to Sol Narwaz. When the police driver went over to a group of rebels with this request, they pointed to one of the nearby damaged shops. Savage and the driver went over to the shop and pushed open the door. A weary-looking Sol Narwaz looked up from his sleeping-bag, giving a wave of recognition.

Five minutes later Peter Savage had introduced a surprised Sol Narwaz to one of his new government officials. Sol did not appear overly impressed and said so, enquiring where this new political leader had come from. It did not help their cause when the rebel government member would not say what part of Libya he came from. All the man would say was: "I am Libyan by birth and anti-Gaddafi; I and two colleagues have been selected by the International Committee of Middle East Advisors to form an interim new government in Benghazi and save Libya from further unnecessary bloodshed."

Former Army Major Sol Narwaz was still not impressed, but asked what this new government wanted from him. "My dear chap, we want you to defeat Gaddafi and capture Tripoli for us." Sol Narwaz's mouth dropped open; he then roared with laughter before managing to splutter out: "What the fucking hell with?" The new government member looked quite upset by the Narwaz taunt, but managed to say: "You have fighters here who will follow you and, as you advance west, more will then join you." Sol Narwaz put his hand on the official's shoulder. "I don't know you and I certainly don't trust you, and that will be the view of these men here. Go back to Benghazi and recruit an army; then we will talk again."

An hour later, after the police driver had prepared a hot meal of rice, spaghetti and tinned fish, the rebel government official who had been silent since the final Narwaz comment spoke. "I would like to return to Benghazi as soon as possible. I must acquaint my two colleagues with the lack of leadership and enthusiasm among the rebel fighters." "We can't travel at night, that would be inviting a disaster," said Savage, adding "We have about two hours of daylight left; I propose we drive into the desert and pick up the tarmac road and drive till dark." And so it was agreed.

They reached the coast road at Sirte at about four o'clock the next afternoon. The sea side of the town was wreathed in black smoke. They joined a gaggle of cars and four-by-fours heading east towards Benghazi, most carrying wounded rebels. The beauty of their current situation was that as long as the convoy carried on, even into the night, it was safe to stay with it. Peter Savage relieved the police driver at 2000hrs and, although the convoy's speed was slow, they were at least moving in the right direction. It was just after midnight when the bright lights of Benghazi came into view, and thirty minutes later they were outside their hotel.

The rebel government official gave a nod of thanks to Savage but not a word to their police driver. He then disappeared into the hotel to find and brief his colleagues; Savage would have liked to have heard the conversation. He waved a "goodnight" to the police driver, who was billeted at the police barracks, and went into the hotel. The first person he saw was Pierre Sarte, gazing moodily into a half-full tankard of beer at a table in the almost-deserted bar. Sarte's face lit up when he saw Savage and he called out: "Please join me, my friend; I have been rejected by your companion." Savage sat down and settled for a coffee rather than a beer and asked if Pierre was fully recovered from his little accident.

"From that injury I am fully recovered," said Sarte "But I have suffered a more serious injury to my pride." Savage sipped his coffee and asked the question by raising his eyebrows. Sarte responded: "Tell me, Peter; is Sally a lesbian? I offered her romance and my body, and was twice refused." Savage decided to play this delicate matter with unaccustomed diplomacy, including perhaps a small falsehood. He took another mouthful of the excellent coffee then said: "I am pretty sure Sally has a boyfriend in

Special Branch, and my information is that they have plans for the future." The effect on Pierre Sarte was dramatic, and he loudly demanded two large brandies from the sleepy barman. His confidence had now returned and, turning to Savage, he winked and said: "I should have realised that when she removed my hand from the top of her leg and wagged a disapproving finger in my face."

An hour later a weary Savage was at last in his room, and a now-content Pierre Sarte was fast asleep in his bed. Deciding to leave a shower till the morning, Savage slipped into bed and within minutes was dead to the world but, in a brief dream, driving his car through what appeared to be a vast field of cotton wool. At six-thirty Savage's travelling alarm-clock started its sleep-piercing bleeping. His practised hand restored the silence and Savage disciplined his senses into some form of order. He shaved and showered, selected a clean shirt and slacks, and made his way down for an early breakfast.

Sally was already there and waved a greeting. The first thing she said as Savage joined her was: "I think I have upset your French friend." Savage smiled and asked: "You rejected his advances then?" Sally nodded. "He is a very attractive man but I do not screw on a first date, or a second for that matter, and when he placed a hand the size of a dinner plate on my upper thigh I put the shutters up." Savage leaned forward in a very confidential manner. "Pierre told me of his rejection when I returned to the hotel last night. I have poured some oil on the troubled waters by suggesting you are shortly to wed someone from Special Branch." Sally laughed before saying: "Thanks, boss, but Special Branch? I have set my sights a lot higher than that bunch of posers."

As the pair tucked into their breakfast, Savage told Sally of their frustrating trip to Misrata and the reaction of the rebel government official to Sol Narwaz's comment. The pair then discussed the death toll in the civilian area as a result of the submarine strike. "I thought an aircraft flying high above the target area as a final monitor made a mistake like that impossible," said Sally. Savage nodded his head before saying: "To me this is another example of some interference with these sophisticated guidance systems; the list of these mishaps grows almost day by day." They were then joined by a beaming Pierre Sarte, who wished them both a bright "Good morning". He then turned to Sally and asked her if she had slept well. "Very well, thank you Pierre; sadly I never had a brother but, if I had, I would have liked him to have been just like you."

In London all eyes were on the Defence Secretary when the morning Cabinet Meeting began. The Prime Minister added to the tension by saying that, when the regular business of the day was dealt with, the Defence Secretary would be making a statement. To those around the Cabinet table the message was clear: that the PM was not shielding his colleague from blame and was, in doing so, distancing himself from criticism. One of the more senior Cabinet Ministers recalled the comment made to him by a respected journalist, who asked: "Is it true you refer to your PM as 'Snow White'?"

The Defence Secretary hardly listened to what his colleagues had to say at the Cabinet Meeting; he was turning over in his mind the statement he was preparing to read. With the business almost over, the Prime Minister looked in his Defence Secretary's direction. The Defence Secretary returned the look, thinking that the two men had once been good friends, but politics and the survival of the fittest had changed all that. He was being left to carry the can for any failure of the PM's Libyan adventure. Although it was not a disaster yet, it was not going as planned and the recent bombing disasters and death toll were not a good omen for the future.

The Defence Secretary, when he spoke, tended to speak at one level. It was as if he was reading every word he said. He informed them that the French submarine Coutance had fired two missiles at carefully-selected targets in Tripoli. A NATO aircraft flying at high altitude took control of the missiles in the last minute of their flight to ensure absolute accuracy. The Defence Secretary paused and looked around the room. Eyes that had once been friendly avoided his gaze. He shrugged and went on, saying that the missiles failed to respond to the guiding aircraft's final commands and both missiles veered away from their target and detonated in civilian areas. He read out the numbers of dead and injured provided by a World Health Organisation source, stating that they were considered reliable. He then tidied his papers and sat back in his chair, his eyes fixed firmly on his former friend's face.

The Prime Minister cleared his throat and spoke. "When we decided on this Libyan action, this was the kind of incident that we most feared, giving our critics just the situation to make capital out of." The Prime Minister then said that, as the missiles were fired by the

French and not ourselves, it was up to them to establish where the fault lay. As the Cabinet Meeting broke up, the Defence Secretary felt a touch on his arm. It was an old friend, now at the Home Office. The man whispered: "I see it was we who decided on the Libyan action now; if Gaddafi is deposed the 'we' will revert to 'I'."

In Benghazi, Peter Savage and Sally Gunn were having a mid-morning coffee with the British Delegate; all three were wondering what their government wanted them to do next. All were agreed that, until a credible rebel government was established with the rebel fighters pursuing a plan, the stalemate would continue. When the British Delegate left their table a friendly-looking American journalist joined them. Savage knew the man quite well; they had met several times in Afghanistan and he knew this man really reported from the front line. Jed Seagrave smiled at Sally and asked her what she thought of this Libyan adventure. Sally laughed before saying: "I think it's a complete and utter balls-up."

Seagrave put three heaped spoons of sugar in his coffee, looked at Sally, and said: "There must be more?" "Oh, there is more," said Sally "Israel bombs the shit out of Gaza and its people, and then commits acts of piracy on aid ships bringing supplies to the people of Gaza from around the world, and the United Nations sticks its head up its arse." Sally took a deep breath and went on: "Now that President Obama and Mr Sarkosy are seeking to be re-elected, and Mr Cameron wants to be seen as a world leader, Colonel Gaddafi is seen as a suitable 'hors de d'oeuvre' and suddenly the United Nations wakes up and joins NATO in this desert adventure."

Jed Seagrave looked at Savage and asked: “I take it, Peter, she is not the official British spokesperson?” “No she isn’t,” said Savage “And you can’t print a word; but you have to admit there are some questions in what she says that demand a reasoned response.” Seagrave smiled and sipped his coffee. “You are so right, Peter my friend, and what prevents us getting these answers are just two words: ‘International politics’. Am I right?”

Chapter Twenty-Four

Several thousand miles away in steamy Penang, Elizabeth Ascott-Gallop was working on the final-year reports of three of her girls. All three were interested in a career in banking and had already had a two-month work experience with HSBC in Kuala Lumpur. One of the girls was going to Hong Kong, where her uncle was a director of The Bank of China; the other two were intending to try their luck in London. Betty AG, as she was known to her friends, remembered the offer made to her by her old friend Carol Andrews-Dixon. Her friend had said she could offer the girls first-class accommodation in return for some clerical assistance in her business. This would give the girls a base to work from while they found the best banking firms to approach.

Betty AG knew that Carol ran a high-class brothel, which was rather like a top-class and exclusive country club. It had a carefully vetted membership, no showbiz rubbish and she had a waiting list longer than that of the

MCC. Among her valued clients were Lords and two Ladies, five Bishops (all Anglican) and a selection of senior Members of Parliament, not all Conservatives. There was also a special list of senior policemen and some shady but attractive figures from the Intelligence Service. From time to time, Carol Andrews-Dixon had been asked to give temporary membership to some foreign diplomats. As long as her fees were met she was happy to do this, providing no violence was involved by her guests in their quest for satisfaction.

The health of her girls was given absolute priority and they were seen frequently by a Harley Street doctor, who was also a member. With an eye to the future Carol kept a diary, in which she recorded the more bizarre requests of her clients. Of particular interest were those of former public schoolboys, who enjoyed really strict discipline. This included use of the cane and the various types of handcuffs and restrictive clothing, all harshly applied. Back in Penang Betty AG decided to ring Carol that evening, which would be about lunch-time in London.

In the MI6 Headquarters in London, a discussion was taking place: whether to recall Peter Savage and Sally Gunn from Libya. They wanted him to go to Afghanistan and India; there were several leads there to be explored and possibly a difficult role for him to play in seeking information. A signal was sent to Benghazi (not coded) to bring Sally Gunn home and Savage on to Cairo; the Foreign Office Junior Minister quickly passed the instructions on. Twelve hours later Sally was back in England and Peter Savage was at Cairo airport waiting for his flight to Delhi. Savage was flown first to Karachi, where he enjoyed an overnight stay in a top-class hotel, then on the next morning to Delhi.

At Delhi he took a taxi to the Empress Hotel, one of the finest in the city; on the way, he was certain he was being followed. After booking into his room, Savage showered and changed into a comfortable but smart cream shirt and drill trousers. He lay down on his bed and fell asleep, catching up on the sleep lost during his journey. If there had been a knock on his door he certainly did not hear it. He was suddenly aware of several hands grabbing him and pulling him to the floor. Savage fought back furiously, landing several good punches which stirred his attackers into a more violent assault. Savage was struck a stunning blow to the back of his head and he lost consciousness; as he slumped to the floor he was handcuffed and dragged from the room.

His attackers roughly carried him to a service lift and down to the rear yard of the hotel, where he was thrown into the back of a police van and sat on for the rest of the journey. The main Delhi police station backs on to the appalling City Prison, where life is cheap and a foreigner's life even cheaper. Savage regained consciousness face-down in a dimly-lit and stinking prison cell. The floor was awash with a fetid mix of water, human urine and faeces. He painfully raised himself to the kneeling position and wiped the floor's filth from his mouth and nose. There were two wooden beds in the cell, one either side against the walls. They were raised about six inches from the mess on the floor; each had a single blanket as a gesture of comfort.

As Savage's eyes became accustomed to the gloom, he became aware that the other bed was occupied and the eyes of its occupant were on him. Savage spoke, his voice still affected by a painful blow to the throat. "My name is Peter Savage and I have no idea why I have been arrested and badly beaten." His cellmate replied: "Just call me Ali. Please obey the rules of this cell; shit and piss on the floor at the bottom of your bed then, when the level rises, you will only have the company of your own filth." At that moment the cell door was flung open and two hunks of bread were tossed in. Ali, with the dexterity of an international cricketer, caught them both; the door was then slammed shut without a word being spoken.

Ali carefully examined the two hunks of bread in the dim light, then generously handed Savage the smaller piece of the two. "Eat while it is still fresh, Englishman; your water jug is at the top of the bed. It must last you twenty-four hours. The cell is washed out every three days; when it is, stand on your bed and hold your blanket over your head to keep it dry." The cell was checked every two hours and each time they were made to stand up to be counted. They were able to keep track of the time because the large clock on the tower of the adjoining police station chimed the hour. Somehow, it did not improve their lot.

Just before midday on Savage's third day in the cell, Ali was taken away for interrogation. About an hour later he was carried back to the cell and thrown in, barely conscious. Savage waited for the door to be shut then he carried his dazed cell-mate to his bed. An hour or so later Ali was violently sick then, obviously feeling better, sat on his bed and sipped from his water jug. "Your turn next, Englishman; tell the bastards what they want to know. If not, make it up. I told them I had fucked Sonia Ghandi; they did not like that."

Ten minutes later Savage was taken from the cell, his slowness to respond to the command prompting a savage kick on the leg. The interview room was reached and he was thrown through the door by two of the massive Sikh policeman; then the door was closed. "Coffee, Peter, or a cold drink?" asked his smiling interrogator, adding "My God, you do smell." Savage chose the water; the coffee could have stayed on his breath, arousing the suspicion of his cell-mate. Peter Savage had met his interrogator twice before, the last time in London. His name was Vidi Sharma and he was a senior member of India's Intelligence Corps.

Savage's role was to try and gain the confidence of his cell-mate and obtain the intelligence the Indian authorities wanted. Ali Shaftri had been a double agent working for both India and China; in recent months his loyalty had shifted completely to China. India was concerned that he had passed information to China on its nuclear capability, in particular its plans for a first strike on Pakistan. Vidi Sharma asked Savage what he knew of India's nuclear strike force. Savage sipped the sparkling, clean, iced water then spoke. "Well, Vidi, you have a rocket capable of delivery and a high-speed and high-altitude bomber to do the same thing. I would guess that in an attack on Pakistan you would deliver the nuclear device by your Indira missile."

Vidi poured some more water into Savage's glass, and then asked: "What do you know of our Indira missile?"

"Well," said Savage "It is an updated version of the American Spartan missile with a far superior guidance system; Chinese, I would guess." Vidi Sharma tapped the table with his finger then said: "That is what we want to know, Peter; this Chinese system which we paid a lot of money for was arranged by Ali. Just how good it is, only that bastard Ali Shaftri knows, and he won't say, as yet." Now aware of what was required of him, Savage was to be returned to his cell; but first, his bona fides needed to be confirmed.

The two massive Sikh policemen were called to the interview room and took a firm hold on Savage. With a shrug of his shoulders suggesting sympathy, Vidi punched Savage twice in the face, causing a cut above the right eye and a deluge of blood from his damaged nose. He was taken back to the shared cell, the door was slammed open and Savage was thrown head-first into the cell's filth. Savage lay still, trying not to allow any of the filthy water into his mouth. He was then helped to his feet by his cell-mate and guided to his bed board. Even in the dim light Ali could see the damage to Savage's face. "I take it they were not happy with your answers, Englishman; another session like that and your own mother would have to look twice before grabbing you to her breast."

The next morning, while it was still barely light, Ali was dragged from the cell for another interrogation. An hour passed and then two. Suddenly the cell door was slammed open and Ali was kicked repeatedly as he crawled back to his bed board. As soon as the door was shut, Savage went to his cell-mate's side; the man was having difficulty in breathing. Savage made him as comfortable as he could. There was something about the man's condition that gave Savage cause for alarm. Ali spoke in barely a whisper; Savage had to bend close to the man's mouth to hear what he was trying to say. He could just make out what he was saying. "I'll share the secret with you, Englishman, but not with those sick bastards. If they want to nuke Pakistan they are in for a shock, because it will never get there."

Ten minutes passed before Ali could muster the strength to speak again. He pulled Savage to him and whispered: "The guidance system of the rocket will reverse course when it recognises any trace of Pakistani territory from the Global Positioning System and it will eventually crash back in India. No nuclear explosion, because the system will not arm itself over the Indian continent." Ali was silent for a moment then asked for water. He spoke again: "It is China's way of preventing a nuclear war which could spread to them." Ali struggled for breath then whispered: "The day is ending early, Englishman, and becoming very dark." Savage eased the man down onto his bed board; the strained breathing grew fainter then stopped. Savage knew at once he was dead.

Savage sat still for a few minutes. It was just another dirty job for him but he felt strangely moved by this man's death. Well, he had the information the Indian authorities wanted; it was now just a case of passing it on, then a good clean-up and a return to a decent hotel. Savage banged on the cell door with his metal jug. A minute or so later it was opened by the two huge Sikh policemen. "He's dead," said Savage. The policemen nodded and dragged the body from the cell then, as

Savage stepped outside, they slammed the door shut again then led Savage up to the interview room. "He did not survive your last interview," said Savage. Vidi Sharma shrugged his shoulders then asked: "What did you manage to get from him before he passed on?" "He could barely speak, but he indicated he had told you fuck-all. In his dying breath he said the Chinese will be pleased, and those were his last words."

Chapter Twenty-Five

As Savage had passed on to Vidi Sharma the last words of the prisoner known as Ali, he was conscious of the intense stare of the special agent. It was obvious and understandable that Sharma had wished for far more but, as far as Peter Savage was concerned, that was all he was getting. Vidi Sharma smiled then stood up and shook Savage's hand. "We must let the doctor clean up your injuries before you go back to your hotel. And thank you for your help, it will not be forgotten." Savage spent fifteen minutes with the prison Medical Officer. Not much was said as the doctor placed two stitches in the cut over the eye and suggested an ice-pack be applied to his swollen nose. As a final thought, the doctor suggested an immediate shower and the liberal use of some antiseptic soap.

A police car returned Savage to the Empress hotel, where the staff made a great fuss over his safe return. Savage contacted the local British mission to find out what London wanted him to do or where to go next; he then crashed out on his bed. He woke up at five in the early evening with a splitting headache and his limbs feeling that they had little or no movement in the joints. Savage used his room phone to speak to reception and asked if a doctor could visit him. Clearly the hotel had been told to take special care of their guest because within ten minutes a lady doctor was at his door. At his request she came into his room with one of the managers whispering urgently in her ear.

The doctor nodded to the manager, who quickly left. She smiled at Savage and said: “It appears, Mr Savage, you are an honoured guest in our country and I am to spare no expense in your treatment.” She quickly stripped off Savage’s shirt and pulled off his tan slacks; she then examined every cut and bruise on his body. His temperature was taken and his breathing checked, as were his leg- and arm-joints. Having finished her examination, she sat on his bedside and gave him a curious look. “Well, my country’s honoured guest, either you have graced one of our penal establishments or been to a very wild party. I suspect the former; what do you say?” Savage nodded, admitting to the penal establishment. The doctor shrugged and said: “You have jail fever, my adventurous friend, plus a normal body’s reaction to some violent treatment.”

She wrote rapidly in her note-book and placed a hand on that of Peter Savage. “You are to stay in your bed here for forty-eight hours; the hotel staff will bring your meals to your room and some medication I will prescribe. I will come and see you tomorrow; just get plenty of sleep.” Savage was just dozing off when a waitress quietly came into his room with some tablets and a bottle of amber-coloured medicine. "Two tablets now, Sir, and a spoonful of the medicine; later I will bring up your evening meal. The doctor has suggested scrambled eggs, followed by some jelly and ice cream." Savage fell asleep almost immediately after the girl had left the room; the dreams that followed were not pleasant.

Savage woke with a start, conscious of something cold touching his forehead. He quickly realised it was the young girl again; she was wiping the sweat from his body, her touch so light it was like a summer breeze. Having finished his head, arms and chest she started on his bare legs. Savage stopped her as, clad in just his boxer shorts, he was becoming aroused. The girl smiled and went into the bathroom to empty the bowl of water; she then poured him a cold drink and, as she left the room, said: "Supper in one hour, Sir." When the waitress returned in one hour, Savage was once more in a deep, but this time, dreamless sleep. The young girl helped Savage into a comfortable position in bed then helped him with his food. He tried to feed himself but his arm shook so badly he would have spilled his supper all over the bed.

The scrambled egg was delicious and then the jelly and ice cream; he finished every mouthful and felt very much better. As the girl left the room she smiled and said: "I will see you tomorrow, Sir, sleep well." Savage was still awake at ten o'clock when a young, very thin man entered the room with his night flask of cold drink. The young man said nothing, just smiled, put down the flask and hurried from the room. Savage took two more of his tablets and the amber medicine then settled down for the night, hoping for a dreamless sleep. At about four in the morning a storm hit the city with great flashes of lightning and enormous claps of thunder. It lasted for about an hour then it moved further east and Savage dozed off again. At about half-past six he woke up feeling much better, so much so that he carefully made his way into the bathroom and shaved then showered.

At eight o'clock the young waitress arrived with his breakfast. He was able to feed himself and the waitress observed that the doctor would be pleased with his progress. At around nine, a messenger brought a sealed letter to his room; it was from the British Mission and his instructions were to return home to London without delay. He was busy packing his suitcase when the doctor arrived. She looked at the suitcase and raised an eyebrow when Savage told her he was required back in London. She sat on the bed and asked him how he proposed to get to London. "Oh, by the midday flight of Air India." The doctor took out her note-pad, wrote something and returned it to her case. "You are not fit to fly, Mr Savage, and I forbid it," she exclaimed.

"How can you stop me, lock me in my room?" Savage argued. "No," said the doctor "I will just ring the airline and issue you with a medical ban on travelling that will keep you in India until I rescind it." Savage spread his arms wide, acknowledging defeat, and enquired: "How about tomorrow?" The doctor smiled. "I will see you at nine in the morning; if there is further improvement you can fly out on the noon flight." Shortly after the doctor had left, Savage had a phone call. It was clearly from 'Six', though that was not mentioned. Clearly they had been informed of his 'illness' and he was told not to rush back.

The day slowly dragged by. All his meals were served in his room by the delightful waitress. Savage allowed himself to fantasise about the colour of her underwear, if any. He watched an English channel on the television; they were showing the very first series of Last of the Summer Wine and he fell asleep. Another waitress brought him his evening meal; she was plump and wobbled as she served his meal. She told him the other girl was off-duty. That night he slept well and was convinced the doctor would allow him to fly back to England. The next morning he was just finishing breakfast when she arrived. After a thorough examination, she seemed satisfied and gave his return flight to England her blessing. Savage thanked her for her treatment and concern and they gravely shook hands.

He finished packing and informed the manager he was leaving, and asked him to thank the staff for their kindness. At ten forty-five a police car arrived to take him to the airport, where he was spared the normal hassle of booking in. When he boarded the Air India plane he was led to Business Class, which was less than half-full. The plane took off on time and Savage slept for most of the journey. The food and service were quite excellent and the two hostesses would have graced any Hollywood movie. They landed in the mid-afternoon at Heathrow; it was, of course, raining. As Savage collected his case and moved into the arrivals hall he saw Sally Gunn. He gave her a brotherly kiss on the cheek and she observed his facial damage. A frown crossed her face and she said: "What does the other guy look like?"

Sally drove him back to his accommodation at MI6 then they went down to the staff restaurant and enjoyed some tea and scones. Sally then left, saying she would see him in the morning. Another agent, very close to retirement, joined Savage from the next table and, referring to Sally, said: "If only they stayed like that." Savage smiled and asked what he meant. The elderly man shrugged and said: "In twenty years' time those firm breasts when unsupported will reach her waist, and that neat, compact arse will be attempting to reach the ground." Savage burst out laughing. "My God, Mac; hardly the view of a happily-married man." Mac sniffed. "My wife was once like that; we are now both in our early sixties, so I speak from experience."

The next morning when Savage went to his office, Madge greeted him with some concern. Clearly, Sally had passed on the details of his accident. Madge had made an entry in his diary that Ralph Maitland would be along to see him at ten for a debriefing. Sally arrived next and she and Madge made some coffee while whispering about Peter Savage's bruised nose and forehead. Ralph Maitland, the former Naval Captain, arrived early and Madge discreetly left the room, leaving Sally to hear about the goings-on in Delhi Prison. Maitland enquired about Savage's health and asked if the cell conditions were as bad as he had heard. Savage said they were, and then described the days and nights he had spent there.

Ralph Maitland interrupted once, when Savage tried to skate over his own injuries when he had met with Vidi Sharma. "Do you think it was necessary for him to beat you as he did?" Savage shrugged. "Well, for my cell mate Ali to be convinced I was not a plant, I think it was." Savage then went on to describe Ali's return from his final interrogation and what he revealed in his last moments of life. Ralph Maitland stroked his chin as he absorbed the intelligence then asked: "But you did not pass that on to Vidi Sharma?" Savage shook his head. "No, Sir; all I told Sharma was that Ali said the Chinese would be pleased."

"So," said Maitland "Any missiles fired at Pakistan by India will not reach their targets, but will reverse course at the Pakistani border due to the way the Chinese guidance systems have been designed." Savage nodded. "It's quite clever really, Sir but, should Pakistan attack India and cross the border, the systems would be effective because the Global Positioning System numbers would relate to India and not Pakistan." Sally poured out Ralph Maitland another cup of coffee. He nodded his thanks and, taking a sip, he said: "We recently have updated our submarine missile guidance systems; they are, of course, our only method of delivering our nuclear weapons." Savage smiled. "I take it these new systems are not Chinese?" "I bloody well hope not," said Maitland "To the best of my knowledge they are a joint venture between Great Britain and America."

Sally Gunn spoke: "I am not an expert in these matters, gentlemen but, if all these Global Positioning Systems rely on a satellite to be effective, what happens if one side destroys the satellites?" Maitland glanced at Peter Savage and sighed. "Tell her, Peter; but remember, the answer is Top Secret." Savage paused, and then took Sally by the hand. "Your life is forfeit if you betray this confidence. If the satellites go down, we revert to the bow and arrow."

Chapter Twenty-Six

At ten o'clock the next morning, at the Ministry of Defence, the three Armed Forces heads met for their weekly briefing. The traditional rivalry between the services was still there, but the bitterness was almost a thing of the past. The senior civil servant who chaired the meeting handed out the information obtained by MI6 regarding the fault in the Indian missiles' guidance systems. Five minutes passed as the senior men of the Navy, Army and Air Force digested the information. The first to speak was Air Vice Marshal Rex Hood, who turned to the civil service Chairman and asked: "Is this for real, or is someone making the best of a failed mission?" Sir Cedric Mason coughed. "The information is completely accurate; our man shared a cell in Delhi prison with the informant, who passed this information on as he lay dying after his final interrogation by the Indian police."

"So the Indian authorities have this information as well then," said Lieutenant General Sir George Walsh-

Spencer. "No," replied Sir Cedric "Our man merely passed on an irrelevant tit-bit. You should also know he suffered a considerable beating to convince his cell-mate he was genuine." The Admiral spoke for the first time. "What is our chap's background? A former serviceman I would guess, to accept a beating in those circumstances." Sir Cedric turned over the paper he had before him, paused and said: "Oh yes, a former Major in the Royal Marines."

Another group of people who found the information Savage had returned with of great interest were based at Filton, near Bristol. Their secure place of work was the original hangar in which the first Concorde was developed and constructed. What they were now working on flew as well, but as the new and highly-secret, intricate guidance system for Great Britain's only nuclear weapons. This was also reliant on satellites for navigation, but not completely. This remarkable piece of miniaturisation could also be radar-guided, identifying the route to its destination by the earth's contours. The hangar was closely guarded by members of the Military Police. It had a secure, double, barbed-wire fence fifteen feet high around the hangar, with additional floodlights and television cameras. The curious Filton locals had been satisfied by some planted gossip that the hangar contained tons of heroin seized in raids on ships in Bristol docks arriving from South America.

Marcia Gibbs was the Chief Scientist in charge and she ruled her ten-person team with a rod of iron. The team's six women and four men had been on the project for just over a year. Originally, the size of the guidance system had been eighteen inches long and twelve inches in height and width; they had reduced it now to just four inches square. At her last visit to the Ministry of Defence, Marcia Gibbs had proudly revealed the four-inch module but, to her horror, she was asked to reduce it further as a planned back-up system was thought necessary. She had been accompanied by her deputy and long-term lover, Helen Palmer. After the meeting they had spent the night together at a first-class hotel in each other's arms, both agreeing what ungrateful bastards men were.

The existence of a new missile guidance system, which should be available to the British military in about twelve months, was not the closely-guarded secret it was thought to be. In London was a Chinese businessman who specialised in obtaining the unobtainable. Lee Peng and his companion, Su Yang, were enjoying the luxury of a penthouse suite in one of London's top hotels. Su Yang had been Peng's mistress for over two years. Recently their relationship had changed and they were planning to marry. This particular evening, Lee Peng was meeting in a bar at Paddington Station a fellow-countryman who worked for the Avon Fire Department in Bristol. The Leading Fireman was known as Li Chang; he was married to an English girl and had been with the fire brigade for five years.

Lee Peng made sure he was at the bar early; he selected a corner table with a clear view of the entrance. He recognised Li Chang immediately and waved him to the table. The two men shook hands and were soon in an urgent but quiet discussion. Twenty minutes later all was settled; the final item had been about money. Li Chang pocketed a thousand pounds, with the promise of a further five thousand when the item under discussion was handed over. The two men shook hands and Li Chang left to catch the next train back to Bristol. Lee Peng collected a double whisky and spent the next thirty minutes watching three provocatively-dressed young ladies trying to lure potential customers away from the bar. An hour or so later he returned to the penthouse suite and the smiling Su Yang. Within minutes, they were eagerly preparing for bed.

The next day Lee Peng had an hour-long meeting with two former members of the Chinese military who were in England, accompanied by their wives, on a tour of London. The tour was paid for by the Anglo Chinese Society, now based in Hong Kong but a front for the Chinese Secret Service. A plan was now finalised in the attempt to seize one of the new missile guidance systems from the secure hangar at Filton. The two former soldiers would ignite part of the hangar roof; once the fire had taken hold the fire brigade would be called. In the confusion their Leading Fireman would do his best to secure one of the new systems from the workshops test-bed.

Almost a week later, at two o'clock in the morning, the two Chinese agents made their way through the deserted grounds of Filton Technical College. At a point on the high ground they were just one hundred yards from the rear of the hangar. In front of them was the boundary fence between the Technical College and the airfield. One of the men carried a pipe-like object which was a lightweight mortar-bomb launcher; the other carried two incendiary mortar rounds. With great care and a good deal of professional confidence they prepared to fire the mortar. Its dull thud-like noise on firing would be largely masked by the busy nearby dual carriageway that twenty-four hours a day carried heavy vehicles to the north and south.

Four Military Policemen were on duty that night, two inside the hangar and two patrolling the double wire fence which secured its perimeter. One of the outside patrols heard the dull thud of the mortar's discharge and turned to his companion and said: "What was that?" Before his companion could reply, a thump and then a series of loud popping noises came from the hangar roof, and the night sky above the hangar was bathed in a brilliant, white light. Inside the hangar lumps of burning phosphorus were falling to the workshop floor, filling the hangar with a noxious, dense, white smoke. One of the two Military Policemen inside the hangar activated the fire alarm system connected directly to Fire Brigade Headquarters and then, in the choking, white smoke, they were forced to flee the hangar and join their companions outside.

The two-man mortar crew were already heading back into Bristol in their hire car, satisfied they had completed their part in the operation. They exchanged amused glances as two fire engines raced by on the other side of the dual carriageway, their lights and sirens not only clearing their way but also waking the local residents. Leading Fireman Chang's appliance was the first through into the hangar compound and within a minute Chang was in the hangar with his two colleagues, protected by their special clothing and breathing apparatus. He directed his two men to check for anything still burning, and commenced his individual search.

Ten minutes later the drama was over; all the doors and windows had been opened and the smoke now almost gone. A team directed by the Brigade's Senior Officer had checked the roof and had covered the ten-inch round holes with a waterproof membrane. The four Military Policemen stood nearby, powerless to stop the firemen seeking the fire's cause. An hour later, all but the fire investigating team had left and returned to their stations. Leading Fireman Chang placed his souvenir of the incident in his personal holdall on top of his locker.

Marcia Gibbs was woken by the urgent sound of her telephone and informed of the hangar fire. She replaced the phone after saying she would come to the hangar straight away. Her companion yawned and said: "Who was it?" Marcia placed a gentle hand on her lover, insisting she stayed where she was. As Marcia Gibbs drove quickly down the A38 from Falfield, where she lived in the old post office, her thoughts were on the cause of the fire. She had already convinced herself that the blame lay with the bloody Military Police being careless with their 'dog-ends'. At the entrance to the hangar she was met by a Military Police Officer who was talking to one of his Sergeants.

"I suppose your men will deny smoking on duty, Captain, but how else would a fire start?" The Captain, who was about to offer Marcia Gibbs a cigarette, quickly slipped the packet back into his pocket. He then said: "My men do not smoke on duty, Miss Gibbs, and the fire was caused by an incendiary device landing on the roof and releasing a phosphorus-like compound, which burned through the roof and fell to the hangar floor, filling the place with smoke." A cold stab of fear caused

Marcia Gibbs to gasp. The Military Police Captain asked if she was unwell. Marcia responded by saying: "Who, apart from your men, has been in the hangar?" The Captain turned to the Sergeant, who checked his list. "Eight firemen; three were in breathing apparatus, the others were just opening windows to clear the smoke."

Marcia Gibbs went into the hangar, dreading what may be missing. Twenty minutes later, she emerged, her face white with shock. She went straight to the Military Police Captain, struggling to control herself. "Seal the hangar, Captain; it is a crime scene. I need to speak to the local police and then London." Thirty minutes later two incredibly scruffy plain-clothes policemen arrived. One, who was a Sergeant, stank of beer; the other seemed to be having trouble with an over-tight pair of underpants. The Sergeant scratched his backside before asking the MP Captain: "What's fucking missing?" Marcia Gibbs gave the Sergeant a look of total contempt before asking: "Are you local Special Branch? That is what I asked for."

The police Sergeant looked at Marcia Gibbs then turned to the Military Police Captain and asked: "Who is she?" Marcia Gibbs spoke before the Captain could reply. "My name is Marcia Gibbs and I am the Chief Scientist in charge of this highly-secret project. I need a senior Special Branch Officer to take charge here and a forensic team to seek evidence. Please arrange that and then return to your bed." The police Sergeant was about to reply, but thought better of it and moved away and began speaking rapidly into his mobile phone. Marcia Gibbs walked slowly into the hangar and sat at a work-bench, at last giving way to a flood of tears.

In London the Duty Officer at MI6 queried the request to provide assistance, suggesting it was a matter for MI5 as it was not an overseas matter. Being that the call had come direct from the Home Secretary's night office, he quickly changed his mind and phoned Ralph Maitland. Maitland growled "Phone Savage, it's his case," then turned over and went back to sleep, hoping to dream that we still had a large Navy. Savage responded to the subsequent phone call with some relief. He had removed the stitches from above his eye before going to bed, and the resulting discharge had stuck his forehead to the pillow. It was now five o'clock and, with some delight, he rang Sally Gunn and arranged to meet her for an early breakfast in the canteen.

It was almost five forty-five when she walked through the door of the canteen. Savage had disposed of his porridge and was about to set about a delicious-looking pair of kippers. Despite the hour, Sally looked very attractive. Savage was about to say so when he thought better of it. "What's all the panic about?" said Sally, trying to find a strawberry in her strawberry breakfast cereal. "Oh, something has gone missing at Filton, Bristol; you know, where they built the Concorde." "Oh, yes," said Sally "My grandfather just about remembers that."

Chapter Twenty-Seven

Just after seven that morning, as Peter Savage and Sally Gunn were starting their journey to Bristol, Leading Fireman Li Chang was going off duty with the rest of the night shift. As Chang drove away from the Fire Station he had an errand to run before going home and hopefully catching his wife still in bed. In the docks were a number of foreign ships. Chang was interested in just one, The Sea Ranger, which was registered in Hong Kong. Li Chang went aboard and handed over his package to one of the ship's officers. He knew that by mid-morning the ship would be in the English Channel heading south towards Gibraltar, and then on to Cape Town. He then drove straight to his home in Eastville and his young wife.

Peter Savage and Sally Gunn arrived at Filton just before ten o'clock. Sally's car had been chosen for the journey because the Honda Jazz was new and comfortable. After at least three pass-checks they finally gained access to the hangar and were confronted by a senior, local Special Branch Officer. After a few suspicious minutes, information was being freely given and they were introduced to a haggard-looking Marcia Gibbs. Gibbs was the only one of her team there; the rest had been told to stay away. "What has been taken, then?" enquired Savage, though he already had a good idea and, of course, Special Branch and MI6 would not be involved in some petty theft. "It's a guidance system; we were in the process of miniaturising it and carrying out tests."

The Special Branch Officer then briefed Sally and Peter Savage on the sequence of events. "Suspects?" asked Savage. "Well," said the Special Branch man "We have the four Military Policemen and eight firemen who were in the building at around the time of the theft; only the four Military Police have been vetted." "Are we are sure that the guidance system was here last night to be taken?" asked Sally. "Well," said the Special Branch Officer "I have been assured by the Chief Scientist that it was." Savage put the same question to Marcia Gibbs, who was sitting alone in the almost-deserted hangar. "Of course it was; we had it on the test-bed and we planned to run a series of tests on it today; all my colleagues would vouch for that."

The Forensic Team had come up with one interesting clue; a fireman's boot-print made after he had walked through the incendiary powder on the floor of the hangar was next to the test-bed. It had come from a size seven boot, rather small by today's sizes. Special Branch had asked for the service records of all the firemen who had attended the call-out. Avon Fire Brigade had at first denied access, after objections from the Fire Brigade Union; then they referred the matter to their solicitors. In any case, within two hours a Judge at Bristol Crown Court had approved a seizure warrant and, after a lunch in the police canteen, the Special Branch team and the two MI6 operatives were ploughing through the records, with particular interest in a fireman with a size seven foot.

It was quickly established that only one man at the scene that night required a boot that size. Li Chang took a shoe size six and a half but the fireman's protective boot did not come in half-sizes so he wore the size seven. It was decided that Special Branch would speak to Li Chang when he reported for his night shift that evening. Peter Savage wanted to speak to Marcia Gibbs about the missing system, so he and Sally Gunn went back to Filton. They found Miss Gibbs sitting in the office sited in the corner of the huge hangar. She looked up from the desk as they stood in the doorway then, putting down her pen, gave a deep sigh before asking quietly: "What is it now?"

Savage came straight to the point. "Miss Gibbs, is your new missile guidance system programmed by a 'memory stick'?" Marcia Gibbs nodded then said: "Yes, the attack program is finalised in the War Room computer then transferred onto a memory stick or sticks in which to programme the missiles' guidance systems." Savage looked at Sally then asked another question. "Has your new system been tested in flight yet, Miss Gibbs?" "Oh yes," said a now more-animated Marcia, "In its larger form before miniaturisation, the missile it was guiding struck its twenty-foot round target spot-on from fifty miles."

After talking to Marcia Gibbs, Savage and Sally made their way to the Fire Station where, shortly, Li Chang would report for duty. They joined their Special Branch colleagues in the canteen there and enjoyed a tasty snack of some beans on toast. They spoke to the Watch Commander when he arrived for his shift, telling him they wished to speak to Leading Fireman Li Chang. The Watch Commander seemed surprised and said: "Young Li Chang is one of my best men and he has been here about five years; he is married to the District Commander's daughter, a lovely girl called Gwen."

Shortly after this Li Chang arrived for duty. He seemed surprised to find Peter Savage, Sally Gunn and two Special Branch Detectives waiting to speak to him. The Special Branch Detective started the ball rolling by asking: "Last night you had a call-out to a hangar fire at Filton. Would you like to tell us about it?" Li Chang looked more comfortable before saying: "Well, it wasn't really a fire. Some device was thrown or launched onto the roof of the hangar; it fell through and scattered some phosphorous-type substance over the hangar floor which produced a great deal of white, dense smoke." Li Chang looked around him and went on: "When we arrived, I and two of my team in protective clothing and breathing apparatus went into the building to deal with any fire; there was nothing burning, just dense smoke, so we opened the windows to clear the air."

"Did you stay together in the hangar?" asked Savage. "No Sir," said Li Chang "We spread out to check for any fire source." Savage tapped the desk then asked: "Did you remove anything from the work-benches in the hangar?" Li Chang stood up, his face red with anger. "No, I bloody well did not; stealing on a call-out quite rightly incurs instant dismissal. I'm a fireman, not a fucking thief." Savage sat back satisfied, now convinced that Li Chang may be a good fireman, but he was an even better actor. Sally asked a couple of questions and there was one more from Special Branch; then Li Chang was thanked and allowed to start his duties.

When Li Chang had left the room Savage turned to his three companions. "Well, I think Mr Chang is our man, but proving it without the missing system and a confession will be very difficult." Sally Gunn nodded her head. "I suspect that the guidance system is a long way from Bristol by now, but a search of his locker here and his car and, of course, his home, may turn up something." The senior of the two Special Branch Officers said: "I'll get a warrant and do his home tonight, and confiscate his mobile phone." As is the normal procedure in these circumstances, Li Chang was released from duty and allowed to accompany the Special Branch detectives to his home. He handed his mobile phone to the detective and asked them not to upset his wife.

Gwen Chang had the shock of her life when the detectives started a search of her home, and sat in an armchair where she was comforted by her husband. After ninety minutes, nothing of any consequence had been found. The final part of the search was concentrated on the Changs' bills and bank statements. It quickly became obvious that Li Chang had two mobile phone accounts; as the police were already in possession of one phone they quickly asked for the other. Chang snapped his fingers in frustration. "Oh, it got nicked from my car when I was paying my petrol bill at a garage." One of the detectives noticed the surprised look that Gwen Chang gave her husband but, for the moment, said nothing.

The Senior Special Branch Detective stood up, ready to leave with his colleague. He smiled and apologised for disturbing Gwen Chang's evening. As they reached the door he turned to Li Chang and said: "Oh, just one more thing, Mr Chang; we need to look at your car." "I'll get the keys," said Chang, his face betraying his annoyance. He returned with the keys but then, after handing them over, went back into the house. The two detectives moved the car into the lighted garage to carry out the search. The first thing they found was the 'stolen' mobile phone in the bottom of the driver's door-pocket. A few minutes later they lifted out the rear passenger seat and found, taped to the underside, an envelope containing paper money. On counting the money, which was in notes of varying values, they found it amounted to a thousand pounds.

They continued their search and took the spare wheel out of the car's boot. Taped to the underside of the wheel was more cash, this time in the sum of five thousand pounds. The Senior Detective instructed his colleague to fetch Li Chang from the house, commenting: "Let's see what the smuglittle bastard has to say about this." On entering the house the detective found Gwen Chang in tears. "He's gone; he has taken our holiday money and his passport, and I was going to tell him that I may be pregnant."

With a full alert out now to arrest Li Chang, and the local train and bus depots under surveillance, Li Chang was less than a mile from his home. He was seated in the rear of a popular Chinese restaurant on a pile of boxes, eating a delicious chicken sandwich. He had already made a phone call to a contact and was told they had tickets to watch Liverpool play. This was a coded message telling him to stay where he was and a car would collect him. An hour later his transport arrived and he was told to lie on the floor in the car's rear by one of the two occupants. Making use of the back roads, the car was driven to Chipping Sodbury and the entrance to a quarry. He was told to get out the car and wait for another vehicle. As Li Chang stumbled out of the rear passenger door a gun was thrust against his head and fired. Chang's lifeless body was dragged by the car's occupants to the quarry's edge and flung into space. Two seconds later there was a dull thud as the body found the quarry floor almost a hundred feet below.

Chang's remains were found ten days later by a man given permission to snare rabbits on the abandoned quarry site. Several thousands of miles further south, the Sea Ranger had called at Cape Town a day previously and the stolen guidance system was now en route by air to Beijing via the diplomatic bag. A still-shocked Gwen Chang had returned to live with her parents, glad now that she was not pregnant, just a few days late. The police and Special Branch enquiry into the missing guidance system was closed and, as no charges had been brought against Mr Li Chang, the money found in his car had been returned to his widow. Peter Savage and Sally Gunn were, however, still very interested and, as Lee Peng and his constant companion Su Yang had been in England recently, the trail of the missing guidance system appeared to lead back to China.

Chapter Twenty-Eight

Great Britain and the United States of America had a long-standing agreement that any serious breach of security to either side should immediately be shared. Great Britain informed the United States of the theft of their guidance system within twelve hours. The news did not particularly excite the Americans; they had a fairly poor opinion of British Security and did not inform their British allies of their progress in this field. In London, Ministry of Defence scientists were struggling to come to terms with the numerous problems of their precision-guided weaponry. They took some satisfaction in knowing that their NATO allies were encountering the same problems, but the pressure on them to find a solution was mounting.

At their base in Southern Italy three Typhoons were preparing for takeoff, their mission to locate and destroy any remnants of Colonel Gaddafi's armour. Their 'eye in the sky' was, of course, the all-seeing AWACS. Three of these huge aircraft, in eight-hour missions, missed very little of any Libyan Army tank or troop movements. Flight Lieutenant Max Samway was leading the flight and so far, despite four missions over Libya, had yet to fire any of his Typhoon's weaponry in anger. The problem was a lack of targets; not surprising, considering the barrage directed Libya's way over the last five months. Now airborne, the three aircraft headed for their 'mission hold' position ten miles off the Libyan coast, hoping it was not just another wasted journey.

At thirty thousand feet the AWACS aircraft scanned several Libyan large towns. Sirte and Misrata were of particular interest because NATO considered the rebels had a good chance of victory there. One of the scanning team in the mission control aircraft reported movement along the road west of Sirte. His team leader took a look and passed the sighting on to his Senior Controller. The Controller digested the information; it was reasonable to presume that anything approaching Sirte from the west was hostile. But was it? He took another look as the light improved slightly, and saw what appeared to be three fast-moving vehicles some three miles west of Sirte, clearly heading for the sea-port town. "That's a confirmed target," he said "Pass it on to the Typhoon Flight Commander."

Max Samway acknowledged his mission, then led his flight at two thousand feet towards Sirte. He spoke to the pilots of Typhoons two and three: "I'll take first shot; if I leave anything, share it between you. Good hunting." The three rebel vehicles were filled with the eager but amateur soldiers, still elated with the success of their overnight raid on one of Gaddafi's desert bunkers. Communications between them and what passed for headquarters in Benghazi were almost non- existent; this all added to the approaching tragedy. All three vehicles had been given an identifying beacon, but with no instructions as to their use. With the lead target now in his sights, Samway fired two missiles and a burst from his heavy machine-guns. The first and second vehicles were completely taken out in Samway's attack; the third vehicle frantically veered left and turned over in the loose sand, presenting an easy target for Typhoon two.

Max Samway reported "Mission Complete" to the controlling aircraft then returned his flight to their hold position. The Typhoons had used half of their fuel by now and were directed back to their base. It was much later in the day when Benghazi complained of an air-strike on its own men. This information was received by NATO Headquarters but was not passed on to the Typhoon Squadron or the Airborne Control Aircraft, and no inquiry took place.

The agreement to provide the Libyan Rebels with supplies by British and French Governments had been deliberately loosely framed. As far as the United Nations were concerned this did not include weapons, but just food and medical supplies. But the French (being French, and President Sarkozy giving a nod rather than the spoken word) dropped weapons and ammunition to the rebels. NATO Commanders said nothing. The problem that became obvious almost immediately was that the sophisticated weapons dropped to the rebels were beyond their limited military skills to use. So now, in complete violation of the United Nations' resolution, France deployed military instructors to train the rebels and assist in their front-line use.

The British Government referred any question by the press about this increased French involvement to the NATO press office. They in turn denied any involvement, claiming that the United Nations' resolution that had triggered the actions against Colonel Gaddafi was being adhered to in word and spirit. In London, the Foreign Secretary requested through his advisors that MI6 agents should be based in Benghazi and send daily situation reports to London, unflavoured by diplomatic caution. It is custom and practice that Ministers never involve themselves directly with branches of the Secret Service, and use an official 'go-between'. The heads of MI5 and MI6 are very wary of direct Ministerial contacts; Ministers are politicians and that is a good enough reason for 5 and 6 to keep their distance.

In his office at MI6 headquarters Peter Savage glanced through the latest daily Libyan Situation Report. The recent 'Blue on Blue' incident near Sirte was mentioned but not explained. His phone rang; it was Ralph Maitland. His instruction was brief: "Come and see me, Peter; I have a job for you." A visit to Maitland's office was always a pleasure, though his ageing secretary guarded his privacy like a Rottweiler. He also employed at his own expense his former Navy steward, who made excellent tea and coffee and provided a delicious assortment of cakes and biscuits. Savage was greeted by Maitland's secretary with a smile, a rare privilege. His steward's greeting was more formal: "Good morning, Major; tea or coffee?"

Maitland got straight to the point: "'Marge' has been on the blower, Peter; apparently those in the mad-house want an unsweetened daily report of what really is going on in Libya." Maitland sipped his coffee and bit into a delicious shortbread. "I have a little gem for you to take, in which to pass your messages; GCHQ have just approved it. Press the 'record' button and speak; when you are finished, press the 'end' button and it will be flashed to a British warship in the Med and they will send it here." A mobile phone, or what looked like a mobile phone, was placed in his hand; it was the size of a packet of ten cigarettes. "Batteries?" enquired Savage. "They will last a lifetime," said Maitland, adding: "Take Sally Gunn with you; it will give you an extra pair of ears and eyes."

When Savage returned to his office he found Sally Gunn talking to Madge. It was obviously about stockings or tights as Sally had her skirt pulled half-way up her thigh. "We have a little job to do amongst the sun and sand," said Savage. "One week or two?" asked Sally. "Two," said Savage "We will start in Benghazi, but I anticipate we will move around as we are required to report on what we find." With the assistance of Madge, a flight was arranged with Air Command, first to Italy and then on to Benghazi, leaving at eight o'clock that evening. They travelled out with some RAF personnel who were beginning a three-month deployment. On arrival in Italy they were told their flight to Benghazi would leave at 0900hrs the next morning. They spent what remained of the night in one of the small hotels near the air base. The hotel owners thought Christmas had come early; they all had a six-month contract with the British plus a bonus for sole use.

The next morning after a substantial breakfast, they flew to Benghazi as the only passengers, having been asked to take charge of a bag of official mail for the British representative there. Benghazi was a little more organised than on their last visit and they were to be accommodated in the same hotel as before. Savage found the room of the Senior British representative and delivered the bag of mail. The welcome he received was hardly friendly, the tubby, half-dressed official saying: "I have heard about you, Savage. I won't stand for any heavy-handed interventions in the very delicate and advanced stage the current negotiations have reached."

Peter Savage made no reply to the stuffy little man and walked quickly away, anxious to obtain some transport for him and Sally to use. As Savage reached the hotel foyer he spotted a familiar figure standing by the reception desk, two large suitcases by his side. "Hello, Pierre; are you just arriving or leaving?" Pierre Sarte shook Savage's hand and then admitted he was on his way back to France. "I had a disagreement with a French Government official and poured a glass of good lager over the fool's head. Unfortunately, he is a personal friend of President Sarkozy, hence the recall." Savage mentioned he was looking for some transport for his stay. Pierre took a set of keys from his pocket. "Take my Honda 4x4, it's almost new and well-sprung."

At that moment Sally came over to them. Savage smiled to himself as Pierre gave her a welcome hug in his huge arms. A few minutes later they were waving the massive figure "Goodbye" as he left to start his journey back to France and, no doubt, a reprimand. At Savage's suggestion the pair made their way to the press area, looking for any familiar face. Sally spotted a TV reporter whom she had met whilst on her crime-busting stint. “Hello, Adam; long time no see. Surely this isn't your normal scene?” “Well, I could say the same about you, Sally Gunn. I suppose you got fed up with the morons in CID seeking a way into your knickers.” Sally introduced Peter Savage to Adam Best and the three sat down and sampled the Press Association coffee.

It was a fruitful encounter. In the situation that existed in Libya at the time, the press normally knew exactly what was going on. But the existing censorship prevented them revealing the true facts, just a sanitised version. The forty minutes spent with Adam Best revealed a great deal- not only was the advance towards Tripoli slowing down, but the Rebel Commanders were at each other's throats. The reason for this was because they all wanted a senior position in the new Libyan transitional government. This was, of course, opposed by the returning and well-educated former Libyan exiles. Also, France and Great Britain had recognised the new Libyan Transitional Council and in doing so allowed them access to the previously-frozen bank accounts which were bursting with oil money.

Of course, as Adam Best remarked, this was not a charitable gesture by Britain and France; they wanted something in return and that, of course, was access to the oil at a very advantageous price and on long-term contracts. Savage and Sally Gunn thought it was time to leave, but an excited babble of conversation started near the entrance to the press office. “Wait here,” said Adam Best “I will find out what all the excitement is about,” and he walked over to the growing crowd by the door.

After a few minutes he returned to their table, his face flushed with excitement and sat down. "Well," said Savage "What's all the fuss about?" Adam Best swallowed hard then spoke: "We have just heard, but it's not yet been confirmed, that the senior rebel commander and two of his aides have been shot dead by other rebel commanders." He paused. "Our informant is sound and I have no doubt the story is true. The British Foreign Secretary will not sleep well tonight."

As Savage and Sally Gunn made their way back to their hotel they were considering what they should do next. Savage spoke first: "I think we should get our kit together and some ration packs and work our way down the coast, to see what effect this piece of news will have on the rebel fighters." Sally nodded, adding: "I take it we will carry our pistols and make sure we have the recognition beacon fitted on our Honda four-by-four?" They decided to make an early start the next morning, and also obtain a travel permit from the Transitional Council.

Chapter Twenty-Nine

Five miles south of Beijing was a well-guarded military complex. It housed an Officers' Training School serving all three services. Almost hidden away in the south-east corner was a special unit, where some of the most inventive brains in China were given a free rein in developing their ideas. A day or so previously, the small unit that Li Chang had taken from the Filton hangar had arrived for the industrial scientists to make their assessment. It took less than an hour for a junior member of the team to work out its purpose. "It is a missile guidance system, clearly freshly miniaturised from an original model. The workmanship is excellent, but it needs to be tested to see if its accuracy matches the quality of its manufacture."

The deputy head of the research establishment took the unit in his latex-gloved hands. It looked impressive, but was it accurate? He turned to his junior colleague. "Set it up on the test-bed and make sure every stage of your test programme is videoed." The deputy head of research then went back to his office and phoned his superior. The conversation was short but the instructions the deputy had been given were clear enough - he was to test the system but was to find it lacked the accuracy of their own guidance system. The team had spent almost twelve months on their model and it must not be found inferior to the one they were about to test.

Twenty-four hours later, and with the test results in his hand, the deputy head waited outside his superior's office. He was called in and waved to a chair. "And what have you found?" enquired his superior. The deputy swallowed hard. "Our initial assessment finds it inaccurate to a worrying degree and vastly inferior to our own new system." His superior stretched his arms above his head. "And that is what your official report will say?" The deputy nodded. After selecting a cigarette from a silver case on his desk and lighting it, the Senior Scientist said: "Now tell me what you really found in this very interesting box of tricks."

Over the next twenty minutes the Deputy did just that, using where appropriate the video recording of the entire test programme; it was stunning. When the demonstration was finished the senior offered his deputy a cigarette, a sure sign he was satisfied. He pointed a finger at his deputy. "Now, my friend, the way is clear. The system we have been given is now our system and, apart from changing the part numbers and English instructions to Chinese, it can go into production and our military chiefs will be delighted." As the deputy turned to go, he was asked one more question. "Is your junior sound, or is it necessary for him to have an accident in the interests of security?" "Oh, he is sound,

Sir; I would stake my life on his loyalty." As his deputy left the room the Senior Scientist blew out a cloud of smoke before quietly saying to himself: "His life or yours matters little, but total secrecy does." He then stubbed out the cigarette.

A few weeks later at the missile-testing range near Lanchow, two medium-range 'Shona' missiles were test-fired fitted with the new Chinese guidance system; their accuracy was stunning. The Head of the secret research centre received a personal letter of congratulation from the Defence Minister and an increase in research budget. The murky world of intelligence was soon buzzing with the success of the new Chinese guidance system, and a king's ransom was promised to any agent who could beg, borrow or steal one. At Filton, Marcia Gibbs was informed of the Chinese success by an old friend from MI5. The news was not unexpected and Marcia's work would continue. It simply meant that two nations possessing nuclear weapons had a guidance system for their missiles equal to one another's.

The United States had a reliable contact at the British Ministry of Defence who passed on the theft at Filton within hours of the Ministry being informed. American Intelligence had highly-paid contacts in China who quickly confirmed that the theft from the research unit at Filton and the successful emergence of a super-accurate missile in China were very closely connected. As a result, the American Ambassador took advantage of a very lively social occasion at the American Embassy to persuade the British Defence Minister in agreeing to share current military research between their two great nations. The substance of the agreement was shared with the Prime

Minister the following day after Cabinet. In the cold light of day, the Defence Minister realised his position was vulnerable. This was confirmed when the Prime Minister remarked: "It's your decision and you are stuck with it. Remember, if things go wrong, you will have to explain your decision to the House."

In Libya, Peter Savage and Sally Gunn were completely unaware of the fate of the Filton guidance system. They had spent ten days moving from one rebel group to another on the pretext of assessing their military needs. They found that the assassination of the top Rebel Commander was what most rebel fighters wanted to talk about. It was an almost even split between the fighters and emphasised that there was more that divided than united them. They came across two poorly-disciplined groups of rebel fighters who had rested up in a coastal village. During their two-day stay they had murdered six men who they said were Gaddafi sympathisers and raped their wives and elder daughters.

When Sally Gunn spoke to the traumatised women it became clear that their husbands had been killed as they tried to protect their families from the rebel fighters' abuse. Peter Savage decided it was as good a time as any to return to Benghazi. Their supplies were almost exhausted, and both he and Sally Gunn were going to make sure that the murder and rape were reported to the international observers. It was early evening as they drove their 4x4 into the hotel car park; one of the hotel staff rushed to help Sally with her kit, whereas Savage was left to manage for himself. As they entered the reception area the large figure of Pierre Sarte rushed forward to welcome Sally with a series of 'brotherly kisses'. Savage had to be satisfied with a handshake, as Sarte explained his return was conditional on a promise of good behaviour.

Later that evening they dined with Pierre and shared their experiences of their latest fact-finding mission along the coast. When they told him of the murder and rape he expressed genuine concern for Sally's safety. She laughed and said: "I'm perfectly safe - Peter is always nearby." Pierre nodded and said softly: "Yes, I suppose he is." They were about to order their sweet when Savage noticed a rather tall, blonde woman standing in the entrance to the dining room, looking somewhat apprehensive. As most of the tables were occupied Savage realised that, with a spare place at their table, she could sit there. He mentioned it to Sally, who nodded and waved for the woman to join them.

Peter Savage and Pierre Sarte stood up as their new companion joined them. "My name is Suki Hodges and I am a doctor from the World Health Organisation. They asked for volunteers, so here I am." Within just a few minutes, Sally Gunn was already deep in conversation with Dr Hodges. Pierre and Savage exchanged glances; their new friend was certainly attractive. She was also very hungry and quickly polished off her starter and main course; then they all started on their strawberrys and ice cream together. It was at the coffee stage that Suki confessed the real reason for coming to Benghazi. She had been recruited to work in Kenya. At the main hospital in Nairobi the senior doctor who was in charge was a Dr Kurt Manheim, who had a predatory attitude to his young, female staff. After he had made two blatant attempts unsuccessfully to lure Dr Hodges to his bed,

Manheim told her that the most unpleasant work at the hospital would fall to her.

Sally Gunn was fascinated; after all, she had experienced almost the same thing in the police. "Where did he send you to work, then?" she asked. Suki Hodges sipped her coffee then spoke. "The bastard sent me to take charge of the 'Rose Cottage' ward." Sally Gunn looked bemused, so Peter Savage completed her education. "Rose Cottage is the internationally-known title for the Venereal Disease Ward, dealing with delightful cases of gonorrhoea and syphilis." Suki laughed. "I was there for three months before this post came up and in that time I suppose I handled more male genitalia than a tart spending a lifetime in a Mumbai brothel."

Pierre Sarte was now so impressed with their new friend that he ordered a round of double brandies. It was just after midnight when they finally went to their rooms, all having had just a touch too much to drink. The next morning when Savage went down to breakfast, he found Dr Hodges and Pierre Sarte half-way through breakfast and getting on like a house on fire. Sally joined them a few minutes later, and it was almost nine o'clock when they all left for their places of work. The special phone that Savage had been given was clearly not working, or rather it appeared not to be. He took it to the British compound and asked one of the technicians to look at it. The man was quite dismissive, saying: "Oh, it's one of those: they cost the earth and are fucking useless."

Sally Gunn and Savage spent most of the morning with the team of International Observers brought in by

NATO, telling them of the murder and rape carried out by the rebels in the small coastal village. Apparently it was one of several the Observers were about to investigate, but they had been told by NATO it was a matter for the United Nations and not them. Later Peter Savage met a reporter he had known some years ago, when British Forces had first become involved in Afghanistan. Over a cold beer, Savage mentioned the murders and rapes carried out by the rebels and the reluctance of NATO to investigate. The reporter drained his glass and then said: "I am not surprised. At our daily press briefing we were told that we should concentrate on investigating allegations of brutality by Gaddafi's men and ignore those made against the rebel forces."

A British destroyer was alongside in Benghazi harbour and Savage took the opportunity to go aboard. He was detained at the top of the gangway by an armed sentry while his request to see the Captain was considered. After he produced his identity card he was allowed to see the Captain who, fortunately, he had met before during his Royal Marine days. Things could not have gone better and he was allowed into the communications room, where he was able to speak directly to Ralph Maitland in London. Before he left the ship the Commander invited him to a mess dinner that evening; it was to celebrate the Captain's birthday and he could bring a guest.

At seven o'clock that evening, Peter Savage and Sally Gunn arrived at the destroyer's guarded gangway; a young, smiling Lieutenant took Sally's arm and led them aboard. Sally was the only female in the Wardroom and enjoyed herself immensely. After dinner, everyone was trying to 'chat her up' and the drink flowed. It was just after midnight when she and Savage managed to escape the party. In the short walk back to their hotel, Sally held on tightly to Savage's arm and stumbled several times. Savage escorted Sally to her room; at the door she kissed him full on the lips, staggered slightly, and said: "Do you want to come in, my handsome friend?" Savage kissed her on the forehead then said: "I would like to very much, but not tonight: we both have had a little too much to drink."

Chapter Thirty

In Tripoli one of Colonel Gaddafi's most trusted aides was completing the damage reports of the last twenty-four hours. Contrary to popular opinion he was not required to hide any damage inflicted by NATO. In any case, most of the targets hit had been hit before and it was becoming obvious to any independent observer that NATO was running out of places to bomb. The Colonel seldom spent more than one night in the same place. He and his advisors worked on the assumption that NATO had people on the ground in Tripoli tasked to report his movements whenever possible.

In recent days the French, through a Russian diplomat, had suggested several ways out of the current impasse. One favoured by several NATO countries was for Libya to be split into east and west provinces and to share the vast oil revenues equally. It was also suggested that both provinces should, within two years of a settlement, have a say on their future form of government. This vote would be available for any person aged eighteen and over and open to both sexes, the vote to be supervised by the United Nations. Britain was becoming a lonely and somewhat isolated voice in opposing some form of settlement. But British oil companies, with their vast investments in Libya, had begun to exert some not-so-subtle pressure in the guise of future political donations.

Most of the world's leaders were now deeply concerned with the financial troubles of America and most Euro Countries. The amount of money being used in the Libyan operation was now being thought of as money not well spent. The Benghazi-located Provisional Government was still viewed with a certain suspicion by other Middle East countries. Egypt, with Mubarak now facing serious charges in the courts, was one of these. They would never forgive their former leader's close relationship with Israel and his willingness to accept the appalling treatment of their Palestinian brothers.

Back in Benghazi,, Pierre Sarte had been approached by one of his country's representatives with a suggestion that if he could visit Tripoli some people there would like to talk to him. France was quite happy for Gaddafi to remain as leader of that part of Libya and was anxious for a settlement with Gaddafi supporters. Pierre, who was making progress in his new friendship with Dr Suki Hodges, was not that keen to go. But he realised that, after his last brush with authority, he was on probation and not in any position to refuse this mission. He decided to discuss this visit to Tripoli with Peter Savage, perhaps even persuading him to accompany him as an observer. After dinner that evening Pierre surprised

Savage with the suggestion. Savage said he would "take advice", leaving the decision to Ralph Maitland.

The next morning he made another visit to the destroyer, HMS Defender and, with the Captain's permission, he again spoke to Ralph Maitland. If Maitland was surprised he did not show it, and told Savage he would have an answer for him in an hour. Thirty minutes later the answer came- he could accompany Sarte, but with no official position and at his own risk. Maitland then added: "Find out what you can, my boy; some people here are more interested than they are prepared to say." When Savage passed on the news to Pierre Sarte he was delighted, and told him that the arrangements were in hand to fly to Tunis the next day and then by road to the Libyan border. Sally Gunn was disappointed not to be included, but was keeping busy trying to persuade the United Nations to investigate the numerous allegations of rape and murder now being levelled at rebel forces.

The flight to Tunis the next day was not without its problems; they were emitting the day's 'friendly forces' signal but were still buzzed by two Typhoon fighters, which made their light aircraft shake, rattle and roll. They were met by a French official at the airport and given the necessary paperwork to allow them both into Libya. Savage was described as English but employed by the French Government as an advisor. They were given a Honda all-terrain vehicle and drove the two hundred miles down the coast to the seaport of Gabes for the night. They were forced to share a two-bedded room for the night, an experience Savage did not want to repeat. Pierre Sarte had settled for a rich seafood evening meal and had developed terrible flatulence, which he made no attempt to contain.

The next morning, after a two-hour drive, they reached the Tunisian-Libyan border. They were carefully checked on the Tunisian side but just waved through by two bored-looking Libyan policemen. They reached the western outskirts of Tripoli just before three o'clock in the afternoon, and booked in without fuss at a seafront hotel popular with western journalists and the International Red Cross. Savage had noticed some bomb damage but not as much as he had anticipated, though the large Police Barracks was now just a burnt-out shell. After an excellent evening meal (clearly supplies were still getting through) they spoke to several journalists, who appeared to be of the opinion that the NATO-led operation was losing its momentum.

During the night Peter Savage woke up when he heard bombs exploding to the east. The raid lasted just a few minutes. He was surprised that there had been no attempt to switch off the street lights; it was as though the Libyan people were saying "here we are - do your worst." The next morning after breakfast, a charming and heavily-built man suggested they went with him to meet certain government officials. If they had hopes of meeting Colonel Gaddafi they were to be disappointed but, in the beautifully-furnished room, a magnificent painting of the Colonel that hung on the wall suggested his presence.

Over coffee and shortbread biscuits the four officials read from a list of demands: 1. The NATO bombing must stop. 2. The Rebel forces should cease offensive operations and return to Benghazi. 3. The United Nations should police Libya and make sure both sides were militarily inactive. The leader of the Libyan group smiled as he finished reading from his list, and awaited their reply. Pierre Sarte looked at Savage, who indicated it was Sarte's call. After a few moments, with his eyes fixed on the room's ornate ceiling, Pierre spoke. "Gentlemen, I think your suggestion should be considered by the United Nations. If you agree, on leaving this room I will pass on your demands." The leader of the Libyan group smiled then leaned forward and touched Sarte's knee. "My friend, we have been in contact with the United Nations. They say it is for NATO to decide, and they are your people."

Pierre Sarte agreed to pass on the demands to NATO, but it had to be through Benghazi. He enquired whether the line was open to that city. The leader of the Libyan group smiled and said: "It is for you, my friend, if you make the call from your hotel." Sarte almost asked the ridiculous question as to whether the line was bugged; of course it would be, he would have done the same. He and Savage discussed the offer as they made their way back to their hotel in the official car. "They won't get a 'yes' to those demands and they know it," said Savage. "It is just a ploy to get some negotiation started." Sarte nodded. "I guessed that, Peter; but it will be your country that will oppose it and, if it comes to a vote, you will get stuffed." Savage laughed. "Oh, I don't doubt it, my friend; my country has fought many battles with yours over the centuries. We get on better as enemies that we do as friends."

An hour later, in Pierre Sarte's room, his telephone was connected to Benghazi. Savage sat on the bed watching the drama unfold. Pierre Sarte spoke for a few minutes, including the Libyan demands. He was clearly asked if his phone was bugged. Sarte laughed and said: "Of course it is, we would do the same." A few minutes later he replaced the phone. "They are going to ring back, Peter, in about an hour. Let's go and get a coffee; I think they are going to suggest a number of intermediate steps before a ceasefire could be called."

In Benghazi the French delegate had been in touch with his government contact in Paris, passing on the three initial demands. The contact, who was a senior government official, had laughed at the Libyan government's clumsy opening salvo, observing: "Well, it's a start, and it shows that at least they have a sense of humour." The Benghazi delegate then asked if their British allies were aware of these negotiations. There was a longish pause then the contact said: "Let us just say they are keeping a listening watch." "And what can I tell our man in Tripoli?" enquired the Benghazi official. After a brief pause, the reply came out in a rush. "Tell your man to suggest to the Libyans that talks could start quite quickly if they agree to them being in Tunis."

Some ninety minutes later, after some technical connecting problems, Pierre Sarte was talking to his Benghazi contact. "I have received a response to the information you provided; please pass on to your Libyan Government contact that they are not in a position to make demands, but France and its allies would readily agree to face-to-face talks that could begin almost at once, but only in Tunis." Pierre Sarte looked at Savage before he replied; Savage, who was on the extension, nodded. "I will pass on your message to our Libyan contact in the morning," said Pierre, and replaced the phone. Both men knew that their Libyan friends had listened in and recorded what had just been said. But the niceties of international diplomacy would be observed the next day when they met, and the look of total surprise on the Libyan faces at the French offer of talks would be quite convincing.

Pierre Sarte and Savage were collected from their hotel at ten-thirty the next morning by a different Libyan official. The man was tall, slim and immaculately dressed and had the air of a highly-successful estate agent. He enquired if they had slept well and then stated that, if they required some discreet and very attractive female company, they had only to ask. Savage immediately altered 'estate agent' in his assessment of the man to 'ponce'. They were taken to the same building and the same room as the previous day. The painting of Colonel Gaddafi had a strange effect on the room - it was almost like being in church.

A general conversation took place before the meeting started, mainly about the money troubles of Europe and America. After some delicious coffee they got down to business. The Libyan delegates looked convincingly surprised at the French offer of talks in Tunis. They then went into a secretive huddle and emerged smiling, stating: "We agree, but we insist you two make the arrangements." Another matter was then raised, which was a request that, once the talks started and as a gesture of goodwill, the bombing of Libya should cease. Pierre Sarte said he would pass on the request to his government contact. They then agreed to meet again the next day, hoping to finalise the arrangements.

Back at the hotel, with arrangements made for the 'unrestricted' phone call at eight o'clock that evening, both Sarte and Savage saw their role there as almost at an end. They arranged for an early-evening meal and then both spent a quiet hour in their rooms, both thinking of what lay ahead. The meal that evening was first-class, and a complimentary bottle of champagne from an anonymous source made it most enjoyable. Then it was back to Pierre's room and the matter of the decisive phone call to Benghazi.

Chapter Thirty-One

The phone call to Benghazi was not as straightforward as Savage and Pierre Sarte had hoped. The main movers in the plan to end the NATO involvement in the Libyan venture had been France, Italy and the Dutch. The British Government wanted the venture to carry on, mainly for domestic political reasons and as a way of pleasing the Americans if the Lockerbie bomber was eventually returned to custody. But in the phone call one thing seemed to dominate the situation and that was the two successes by Rebel Forces in now controlling towns on the main route to Tripoli. Pierre Sarte was told to delay the Tunis talks for ten days, not cancel them and, at that point, the line to Benghazi went dead.

This came as no surprise to Sarte and Savage as they knew the Libyans were monitoring the call. The next morning after a full breakfast they were collected and taken to the meeting point. tThe expressions on their hosts' faces proved without doubt that they had heard all of the previous night's conversation. Their chief spokesman did not bother to hide the fact he was aware of what they had been told to propose. "Gentlemen," he said "I am very disappointed that your countries have decided to delay our talks when we were so close to a breakthrough." Taking a sip of water, he went on: "When you return to Benghazi tell your people that a couple of villages captured by the rebel murderers is not the success that they claim. Contact us when you wish to talk again."

Pierre Sarte and Savage were then driven back to their hotel. There they stayed for lunch then began the drive back to the border with Tunisia. They stayed at Gabes overnight; this time the hotel provided them with separate rooms. The next day after an early start they reached Tunis in mid-afternoon. They reported to the French Consul who, after making sure their Honda vehicle was undamaged, drove them to the airport and their waiting plane. It was the same pilot who had flown them from Benghazi and he gave them both a cheery wave and, with the minimum of fuss, contacted the control tower and was told to proceed.

The pilot kept well out to sea on the flight back to Benghazi, chiefly to avoid the notoriously itchy fingers of the rebel gunners. By six in the evening they were back in their hotel and sharing their story and some tea and coffee with Sally Gunn. Later at dinner that evening the three became four, as Suki Hodges joined them. Clearly Suki and Pierre were becoming quite close and, when Pierre left the table for a few minutes, Sally enquired quietly of Suki if they were now lovers. Suki giggled and whispered: "Not just yet, but I have handled his equipment and it will be a very tight fit." With the two girls now in a state of helpless laughter, Savage enquired "Why?" "Oh it's a woman thing, Peter; you wouldn't understand," said Sally, tears of laughter streaming down her face.

At that point a familiar figure approached their table and Savage scrambled to his feet, a surprised smile on his face. He gave their unexpected guest a brotherly kiss on the cheek and clasped her hand, before announcing to his three companions: "My friends; please allow me to introduce Captain Jane Downey of the Royal Military Police." Savage grabbed a spare chair and placed it between his chair and that of Sally Gunn. Jane turned to Sally and said: "We almost met in Afghanistan, but you were hurt in that explosion in Pakistan; are you now fully recovered?" Sally said she was, and said: "I do not want to appear rude, but what are members of the British Army doing in Benghazi?"

Jane Downey realised that here was one woman who was not offering eternal friendship or a share in Peter Savage. She managed a smile and said: "I am here to work under the United Nations, who want me to investigate a series of rapes that have allegedly been committed by the Rebel Fighters." Peter Savage, who could sense the hostility between the two women, jumped in with both feet before the situation got out of hand. "Well, Jane; things could not have worked out better. Sally here has already collected a great deal of evidence in these cases - you two will be able to work together." The expressions on Jane's and Sally's faces suggested otherwise, but Savage decided to ignore that.

Dinner went better than expected, mainly because Jane Downey and Sally Gunn decided to pretend the other was not there. However, when the coffee and brandy stage was reached the relationship had thawed slightly; but not a lot. Jane was the first to leave the table for bed, claiming that the flight from Afghanistan to Benghazi via Cyprus had been most uncomfortable and she was tired. The next morning, Sally Gunn and Jane Downey were both at the breakfast table when Savage joined them. There was no sign of Suki Hodges or Pierre Sarte, and both Sally and Peter Savage decided not to speculate on their absence. Jane was hoping to make a start on her investigation and, turning to Sally Gunn, said: "I would really appreciate your help in getting this matter off and running Sally; you know the ground and some of the victims. With you involved we can save a great deal of time."

Peter Savage waited for the expected refusal but it never came. Instead, Sally Gunn gave almost a smile then offered her hand to Jane Downey; for the moment, all was well. Savage was about to go and see the British and French delegates when a breathless Pierre Sarte appeared, but there was no sign of Suki Hodges. Savage, curious as to her absence, enquired if Suki was unwell. Pierre Sarte's face took on the look of a guilty schoolboy and he stammered out that she was a late riser. As they were late for their meeting, Pierre had to be satisfied with just one cup of coffee, and then they dashed off. The two delegates were reading through a pile of signals when Savage and Sarte reached them, and they were waved to a chair.

"Did you know that Gaddafi's forces still have Scud missiles as part of their hidden arsenal?" said the French delegate. Savage spoke first. "Well, if they do I would not think they would be very reliable; the Russians were dumping them everywhere a few years ago." The British delegate laughed. "The Gaddafi forces fired one yesterday but it landed harmlessly in the desert; we think they could have about two hundred." Pierre Sarte looked at Savage then said: "I would wager a month's pay that if they have more than twenty operational, that's their limit; and do they have the warheads?"

This same question was being asked at NATO Headquarters, where they had been wondering what targets they had left to bomb. A search through NATO intelligence suggested that the Libyan Scuds were type 'B'; how many were fully operational was anyone's guess. Most intelligence sources suggested between twenty and fifty, and as for warheads, none was nuclear or chemical and the explosive content could extend or reduce the missiles' range. Experts in NATO thought the Libyan missiles would contain between five hundred and eight hundred kilos of high explosive, it being a mix of TNT and RDX. Expert opinion of the Scud did not rate it highly; after all, it was based on the German V2 Rocket from World War II, and was considered inaccurate and unreliable by present-day standards.

There was another matter that was now occupying the top brains in NATO; this was supposing the rebel forces did manage to capture Tripoli. It had always been assumed that some form of cease-fire would be in place before the notoriously badly-disciplined rebel forces got that far. But now, after the advances of the last few days, that may no longer be the case. The answer would be a United Nations force inserted to prevent mass murder, but no such force existed at the moment and could take weeks in the forming. Another matter causing concern was that, since the murder of one of the rebel commanders who had been seen as their leader, who now among the rebels was the 'top dog'?

In Benghazi Peter Savage and Pierre Sarte, having discussed the present situation, were expecting to be sent back to Tunis and then on to Tripoli. Sarte had other matters on his mind since the successful deflowering of the delightful Suki Hodges. It was normal for Pierre to enjoy his success with the ladies briefly, then move on to other temptations. This time it was different, and Pierre was almost convinced he had met the 'chosen one'. A senior French Air Force officer had appeared in the hotel late that morning and Pierre Sarte drew Savage's attention to the new arrival. "I know him, Peter, he is a top operations man; but what is he doing here?" Savage looked at his friend and smiled.

"If you know him, Pierre, why don't you ask him?" Sarte stroked his chin, and then made a decision. "Come with me, Peter; we will do just that."

If Claude Masta was surprised by the sudden appearance of his old friend Pierre Sarte, he did not show it. He shook hands with Sarte, then Peter Savage and then said to Sarte: "Trust you to be where confusion reigns, Pierre; it appears to be the only situation that you are comfortable in." The three ordered coffee, then Pierre Sarte asked Claude Masta the direct question. Masta smiled and sipped his coffee before replying: "I could say that I like the sea, sand and sunshine; and I do, but my job is to stop the rebels entering Tripoli." Savage spoke for the first time. "And how can you do that, my friend?" Claude Masta pointed a finger into the air. "I specialise in air operations. I am here to talk to the rebel commanders and tell them that, when they get to within five miles of Tripoli, they must stop and dig in." "And if they don't?" enquired Savage, almost now guessing the answer. "In that case, they will be subject to NATO airstrikes around the clock until they withdraw back to an agreed line in the sand."

Peter Savage immediately could see faults in this plan. In some ways it was incredibly naive; after all, we are not talking about a well-disciplined rebel force. Their main commanders were always some distance from the fighting, and who decided when and what to attack was often down to a majority decision at the scene. It may well be that the commanders at a distance agreed to the five-mile halt and dig in, but it was another matter to expect the bloody enthusiasm of the rebel fighters to be curbed with a great victory almost in their hands. Both Savage and Sarte made this point with Claude Masta. He shrugged his shoulders and sipped his coffee, seemingly unconcerned. He then touched his lips with his serviette and studied both of his companions' faces, then spoke. "My senior Generals have a similar view of the rebel forces as you have. If they reach Tripoli they will have completed their role." He paused, lit a cigarette and then, through a haze of smoke, said: "And if they fail to halt at five miles we will dispose of them, and that will be another problem solved."

Chapter Thirty-Two

In the United States of America the matter of Libya and the success or failure of the rebel forces was way down the President's list of priorities. One of the President's young advisors mentioned that if Tripoli fell and Abdel Basset Al Megrahi, the freed Lockerbie bomber, was recaptured he could be taken to the United States. There he could be incarcerated for the rest of his life, which would be popular with the American electorate. A glimmer of interest showed momentarily on the President's face, but he had more serious matters that demanded his attention. The young aide scrawled 'later' next to the bomber's name then, a second or so later, added 'much'.

In London earlier that day, the Prime Minister and the majority of his Cabinet colleagues had given up on their holidays and were still deeply engrossed in the previous week's street violence. One or two of the cabinet had decided to continue their rest in the sun, but it was widely-felt that they were due for the chop in the next reshuffle anyway, and they knew it. As in America, Libya was not top of the Government's priorities - it had been overtaken by other events. The Prime Minister was more interested in Libya than his Cabinet colleagues. He had much to lose if things there went badly but, if things went well, his position as a world statesman would be enhanced.

At MI6, Peter Savage's daily report from Benghazi regarding the enforced five-mile limit outside Tripoli was raising a few eyebrows. One matter immediately raised was the presence of large numbers of rebel sympathisers in Tripoli who were prepared to join in the fight once their rebel brothers entered the city. If the entry ban was enforced it could well prolong the siege, something NATO certainly did not want. This information was being shared with the Ministry of Defence; this was no great gesture of solidarity between Government Departments, but ensuring that MI6 was now owed a favour. When the Air Force Board were challenged over the French line on Tripoli, they blustered that it would be French aircraft that would have to enforce it when, in reality, they knew nothing about it.

Jane Downey and Sally Gunn were getting on far better than had been expected. The investigation into the rapes and assaults carried out by the rebels was not going well. Sally had a list of names and the addresses of the victims, obtained very shortly after the assaults took place but, when they went to the addresses, the victims had gone. One elderly woman, who Sally thought was the mother or grandmother of one of the victims, was reluctant to talk to them. After some gentle persuasion from both Sally and Jane, she hinted that money had changed hands and some advice had been offered. "What kind of advice?" asked Jane. The old woman shrugged then said: "They said in times of war much of what happens is best forgotten."

The two women returned to their hotel, both realising that any chance of any rebel fighters facing a court over their brutality was slim indeed. Sally Gunn was surprised that she did not rejoice over the apparently wasted journey of her rival for Peter Savage's attentions. She surprised herself even more by saying to Jane that, if an occupying rebel force were eventually to enter Tripoli, its behaviour would require close supervision. That evening Jane Downey repeated Sally's words to her superior in Afghanistan. There was a pause as he gave the matter some thought; then he said: "In that case, Jane, I suggest you stay in Libya until the situation there is resolved."

When Peter Savage joined the two ladies for dinner they were chatting like old friends. He gave them both a quizzical look then said: "OK, let's have it; this morning you were at each other's throats like a couple of alley-cats. Now you are like a pair of flaming 'Ovaltinies'." Jane looked at Sally before answering, and then said: "We are both in the police business, one military, one civil. Matters could rapidly come to a head in Tripoli and we could be up to our neck in cases there, so we have resolved to work together." Sally nodded and added: "As we both like you, Boss, it seems the right stance to adopt; so rid yourself of that worried frown."

Any tidy, military mind will tell you that working to an agreed plan would allow those involved some leeway on their good and bad days. Conversely, a group of enthusiastic but ill-disciplined and heavily-armed men seldom work to a plan. They have an objective in mind but no strategy on how to reach it as part of a coordinated effort. Twenty-four hours after Jane Downey and Sally Gunn had spoken to Peter Savage, five uncoordinated groups of rebel fighters closed in on Tripoli. The speed of their assault owed much to a record number of air-strikes carried out in the Tripoli area by NATO aircraft. Gaddafi loyalists moved from the outskirts of the city to the safer, narrow streets to avoid the deadly assault of the Apache helicopters and the spying eye of the unmanned Predator aircraft.

All this came as a complete surprise, not only to the NATO advisors in Benghazi, but also to the members of the Provisional Council sitting in safety there. Pierre Sarte and Peter Savage were immediately ordered to Tripoli and, after some thought, Jane Downey and Sally Gunn were added to the list. The journey was to be made by helicopter. The Sea King was on loan to the Red Cross and, apart from Peter Savage and his colleagues, it would carry medical supplies and a Red Cross surgical team. The flight, which would take between two and three hours, was scheduled for seven o'clock the next morning. Night flying would attract fire from both sides and, despite the urgency, it was discounted.

The next morning, as they enjoyed an early breakfast, they were informed that rebel fighters had already entered the city of Tripoli from the east and west. They were making surprisingly good progress, capturing the heavily fortified east gate of the city with hardly a shot being fired. Savage and his team boarded the elderly Sea King on time but were made to wait forty-five minutes for the Red Cross surgical team to arrive. When they finally arrived without a word of apology, one of the surgeons objected to Savage and his colleagues taking passage as they were not Red Cross personnel. Pierre Sarte reacted furiously and would have thrown the surgeon off the helicopter but for Savage's intervention. The chastened surgeon then decided to withdraw his objection.

As the helicopter took off and headed west, Savage shouted in Sarte's ear: "If you should get wounded and have him as your surgeon he will probably start proceedings by removing your balls, just for old time's sake." The flight took twenty minutes longer than scheduled, as they flew a course five miles out to sea after coming under rebel fire from the coast road. They landed on a football pitch which was on the eastern outskirts of Tripoli, and they were immediately surrounded by three machine-gun-mounted pickup trucks. The Red Cross team and their supplies were greeted with a great deal of clapping and shaking of hands. Savage, Sarte and the two girls were viewed with some suspicion after their description of themselves as NATO observers, which was not received with any great enthusiasm. Where they had landed was about a mile from the captured East Gate and, mainly due to Jane's and Sally's charm and their tight-fitting shirts and trousers, they were given a lift there.

Their search for a vehicle was solved in the military car park at the East Gate. A very fat, middle-aged rebel fighter, impressed with their two female companions, threw them a set of keys belonging to a khaki-painted Ford pickup truck. A quick check of the fuel gauge showed that it had recently been topped up and, as it started up straight away, 'Fatso' had done them a favour. With Pierre Sarte driving, they joined a convoy that was making its way down to the port area. Progress was slow and twice the convoy stopped altogether as it came under fire which, on the second occasion, came from a group of rebel fighters driving towards them. None of the rebel groups appeared to have the slightest idea of 'fire discipline' and, of the number of casualties they sustained, almost half could be classed as 'friendly fire'.

Another factor that the world's press was yet to come to terms with was the tribal system in Libya. Colonel Gaddafi over the years had very cleverly played one discontented tribe off against another. Tribal land found to contain vast amounts of natural gas or oil would receive large sums from the government to be spent on housing, hospitals and schools. Other tribal land not blessed with these natural gifts would struggle to educate their young men and women and have to travel miles for hospital treatment.

This discontent among the tribes was well understood by Savage and Pierre Sarte; they had experienced the same tribal rivalry in Afghanistan. Some of the Afghan tribes had land well-suited to the growing of poppies which, of course, converted to opium and on to morphine and heroin, making them rich. The hill tribes struggled to survive on poor land with their goats and sheep, just about keeping starvation from the door. Libya has some one hundred-plus tribes, not all of them anti-Gaddafi, and getting them to build a new society together would be no easy matter. By early afternoon they found themselves sheltering in the former central police barracks as two tanks taking advantage of the narrow streets were holding up one column of the rebel advance.

Later that afternoon the astonishing news reached them that one rebel column had entered Gaddafi's secure compound and was meeting only light resistance. The task had no doubt been made easier by several NATO air-strikes that had destroyed part of the defending three walls. It took Savage, Sarte and the two girls ninety minutes to get there and, by this time, widespread ransacking was taking place of the Colonel's private quarters and the well-stocked armoury. Of Gaddafi there was no sign and, in the disorganised chaos, no attempt was being made to search the suspected vast number of tunnels under the complex.

At seven in the evening they returned to the relative safety of the police barracks to prepare some food and settle down for the night. About a hundred rebel fighters had made the same decision so they had plenty of company and, as it was still Ramadan, as soon as the sun set the rebels began to prepare their meal for the day. Sleep for the rebel fighters, and almost anyone in Tripoli that night, was broken by frequent and sometimes heavy gun-fire. Strangely enough Jane and Sally rarely woke, whereas Peter Savage and Pierre Sarte seemed to hear every shot and shell that was fired.

Just after six o'clock Savage gave up any attempt at sleep and, after giving Pierre a kick up the backside, began to think about breakfast. The rebel fighters were already on the move but it was difficult to see who was leading them. Where they went, and why, appeared to be a collective decision. Savage watched them leave, some of the rebels casting envious glances at the still-sleeping girls. Breakfast was constructed from the United Nations twenty-four-hour ration pack marked "European Forces". With two large pots of coffee bubbling nicely, the two girls were woken and immediately required some privacy for a pee. With the rebel fighters now departed, the burnt-out shell of a three-ton truck satisfied their modesty.

An hour later the four were ready to begin their day exploring Tripoli and wondering whether the next few hours would see the end of Colonel Gaddafi and his forty-odd years of leading the Libyan people. As they drove their vehicle out of the ruined entrance of the police barracks, they found themselves face to face with a massive tank that had just lurched to a halt in a cloud of oily fumes. Its turret slowly turned in their direction and they found themselves looking up into the barrel of its huge main gun.

Chapter Thirty-Three

For some thirty seconds or so nothing happened; then there was a 'clang' as the tank commander's hatch was thrown open. They found themselves staring up at a bearded, grinning face, the owner of which said, almost politely: "Back up, or be fucking flattened." Pierre Sarte, who was driving, 'backed up' and, in a cloud of blue, oily exhaust, the tank lurched along the road. It noisily crashed into another gear with the expertise of a vicar's wife learning to drive very late in life. Whether the tank was Rebel or a Gaddafi supporter was never clear, the uncommitted being at risk from both in these confusing times. Pierre Sarte was intending to visit the Gaddafi compound, which had been captured yesterday, but was turned back and diverted onto the coast road. A few minutes later they were opposite Tripoli harbour, where there was a great deal of activity.

Three ships were in harbour, all busily taking on passengers waving their exit papers, issued by a group of United Nation officials seated just inside the main gate.

At this point they heard the noise of low-flying aircraft, followed by three huge explosions which rocked their vehicle from side to side. Several columns of black smoke began to rise from the south of the city, where sporadic firing had been heard since daybreak. They drove to the football field where their helicopter had deposited them and brewed some coffee on their solid-fuel cooker. Two well-dressed men, looking completely out of place, wandered over to them attracted by the smell of their coffee. They spoke a greeting in French and soon Pierre Sarte was busy extracting from them who they were and what they were doing there.

Sharing a mug of strong coffee, the two men explained they were tasked by the United Nations to visit all the hospitals in the Tripoli area and compile a list of supplies that were required urgently. They had been dropped off by a helicopter from Benghazi, but had been unable to obtain transport to begin their mission. Jane Downey and Sally Gunn exchanged glances and, with a brief word with Savage, walked over to the transport compound where their old friend 'Fatso' was still clearly in charge. After a few minutes conversation the deal was done and both women submitted themselves to a hug and a very wet kiss from their plump friend before driving off in their loaned vehicle, a well-kept Honda HRV.

Within minutes the two French Officials, having expressed their thanks, drove off to begin their task of finding out the emergency needs of Tripoli's five hospitals. Peter Savage and Pierre Sarte were anxious to have the opportunity of seeing for themselves the vast underground complex of tunnels beneath Gaddafi's headquarters. Their second attempt to gain entry to the complex was more successful and, with Pierre shouting a stream of French at every rebel soldier who queried their presence, they arrived at the main building. Once there they were surprised to find that no one queried their intentions. The fact was that, having got this far, the rebels assumed they had permission to be there.

Their two female companions were not that keen on exploring the dank and poorly-lit maze of tunnels beneath the main building, and opted to stay above ground and guard the vehicle. Savage and Sarte, each armed with an automatic pistol and a powerful torch, descended the ladder into the main tunnel. It was not a pleasant experience, as the ladder shaft had been used as a toilet by the capturing rebel fighters. The two men wiped the filth from their hands on the walls of the tunnel then, twenty yards further on, found a latrine with running water and a liquid-soap dispenser. Now feeling much happier and smelling of roses, they pressed on. Some stretches of the passages had light switches that actually worked and they progressed faster than before. The obvious ransacking of the adjoining rooms became less obvious; clearly fewer rebels had come this far. They spent some time in one room which was fitted out with radio equipment, sadly now most of it smashed. Next door was a computer room, which had been selectively looted with an eye to making a quick profit.

A few yards on and they found a massive, metal-faced door with three locking points; some attempt to open the door had been made, without success. The rebel searchers had taken the easier route straight on, but Savage and Sarte wanted to find out what lay beyond the metal door. Savage returned to the computer room and searched through the desk drawers for any keys; a curious Pierre Sarte joined him. Disappointingly, the drawers contained no keys. Savage then pulled out each drawer to see if anything was taped to the undersides. Drawer three was the prize, with three large and shiny keys taped securely to the underside. Pierre muttered his congratulations as the keys were removed, and both men moved to the massive, metal-faced door. Peter Savage took his time in selecting each key to match a lock. After a few minutes all the keys were in their correct position, and he first turned the top key, then the centre key and, finally, the bottom key.

The door failed to open, so he relocked each key point and changed the unlocking sequence to middle, top, and then bottom. With the hiss of a well-oiled unit, the door opened and the air within the new tunnel was like nectar. Clearly, this air-conditioning unit was still working. As this part of the underground complex had not been visited by rebel wreckers both men stood perfectly still for a few minutes, listening for any sound of occupation ahead of them. Two minutes later, now satisfied they moved on, but now with their pistols drawn. The first side-room they came to had six beds, each made up with clean bedding. A few yards further on was a kitchen; it was well-stocked and spotless, with no evidence of recent use.

Next door was a computer room. It had six units, two of which were switched on. One had recently received some messages which the printer had processed. Savage took the top two out of the tray, glanced at them, then passed them over to Pierre; they appeared to be in Arabic.

They moved on, passing two well-furnished, single-bedded rooms. From then on the floor began to slope down. At the bottom of the slope was a set of buffers and the beginning of a narrow-track railway system. Looking down the track, the tunnel was in complete darkness. Both men knew that they had come as far as was possible on this day.

They made their way back along the tunnel, pausing to take all the paperwork still in the printing tray. They relocked the massive, metal door and kept possession of the keys, intending to return the next day. Making their way back up the passage, the stink told them they were getting closer to the filthy, shit-spattered ladder. Taking a deep breath, they climbed the slippery rungs into the fresh air, wiping their hands on some discarded clothing, and then walked over to their transport and their two bored-looking female companions. Sally was the first to respond, wrinkling her nose at the foul smell coming from both men. "Been at the latrines again, have we boss?" Sally enquired. Savage explained about the misuse of the ladder shaft and their subsequent progress through the locked door.

As they drove to the compound's exit they noticed that, apart from the normal rag-tag rebel guard, there were two armed men who were clearly European. They were instructed to stop, and one of the Europeans said in French: "Which one of you two is Sarte?" Pierre replied and there was a rapid exchange in French. Jane Downey, who spoke schoolgirl French, managed to pick up the occasional word, including "sleep" and "tonight". But all was revealed when the rapid conversation finished and Pierre Sarte explained all. Pierre gestured towards the two men. "These two men are French Special Forces. They are very keen to join us tomorrow when we explore our mystery railway. Tonight we all stay at the Argos Hotel, which the Special Forces have commandeered."

With the explanation complete, they drove off to the Argos Hotel with their two extra passengers, both of whom were giving Sally and Jane appreciative glances. At the Argos only the second floor was in use; the top floor had sustained some slight damage during an air-strike. The rooms were clean and the staff friendly. Peter Savage and Pierre took the chance to shower and rid themselves of the unpleasant smell. They all dined together and found that both of their new friends spoke quite good English, and the evening passed agreeably. Peter Savage made use of a phone and contacted London, well aware he could be 'overheard'. His instructions were clear - he could follow any lead his instinct suggested.

The plan for the next day was discussed. Savage, Sarte and the two Special Forces men would carry on where Savage and Sarte had stopped. Sally and Jane would spend the day with an interpreter, trying to make sense of the paperwork found in the computer's print-out. After one brandy each it was off to bed, with an early breakfast already ordered for the four men. Tripoli enjoyed the quietest night since the rebel offensive began, with just the brief rattle of celebratory fire from the direction of the airport. Peter Savage woke just after five o'clock. He lay there thinking of what lay ahead down the dark railway track and what progress they would make. At five-thirty he left his bed and shaved then showered. Pierre Sarte was in the next room and he could hear him on the move.

At six o'clock he went down to the dining room and joined his new French friends. Pierre was last, as usual. Then they all enjoyed their full breakfast. At six forty-five they drove to the compound, not too early as the half light could be dangerous with an army of poorly-trained men who tended to open fire without a moment's thought. One of the French Special Forces men spoke to one of the compound gate guards to make sure their vehicle was kept safe. All four men moved over to the main building and the entrance to the tunnel. Savage and Sarte carried their automatic pistols; the two Special Forces men had compact, light machine-guns. They all carried freshly-charged, powerful torches, a long, dark and unfriendly tunnel being no place for a failing light-source.

The descent into the tunnel down the evil-smelling and slippery ladder was made without comment. Sarte had briefed his compatriots that a wash-place was but a few yards away. Once they were clean again, they moved on to the massive, locked door. There, Savage produced the keys and then went through the unlocking sequence. Their two companions were impressed as the door that must have weighed over a ton opened with just a whisper. The four adventurers moved along the passage, with the two French Commandos determined to take the lead. They paused at the computer room. Savage noticed that at least a further ten sheets of A4 were in the printer's tray. Another few minutes and they started down the slope to the railway line; it all seemed to be as it was yesterday afternoon.

They began to move along the railway line, their torches searching for any tripwires or anything unusual.

The four men kept at least five yards apart in the interests of their security. Every fifty yards or so, they stopped for a minute, listening for any movement ahead. It was all suspiciously quiet. Ten minutes later they could see a light ahead, and they switched off all their torches and moved with even greater care. Their caution proved sensible, as they could now make out the sound of several voices less than eighty yards ahead.

Chapter Thirty-Four

Taking great care, the four men moved forward another twenty yards. The light ahead suddenly dimmed and they heard the whine of an electric motor. Then the vibration from the tracked vehicle moved up through their bodies as the train and its occupants moved off at speed. Within seconds they were again in complete darkness until they switched on their torches, the noise of the departing train barely noticeable. Two minutes later they reached the position the train had occupied; it was a lay-by. But this was no ordinary lay-by; on the raised concrete platform was a huge pile of boxes. What they contained was stencilled on their sides in red paint; it indicated their contents were explosives. The ten-layered mound also had another unwelcome surprise; two wires made their way into each layer from a control box that had a light pulsing, indicating it was active.

One of the French Commandos spoke quickly to Pierre Sarte after he had briefly examined the control box. Pierre Sarte turned to Peter Savage. "We can take our time making this lot safe - the twenty-four-hour timing clock is set for midnight." "That's good news," said Savage "But have the bastards installed any failsafe tricks?" The two French Commandos exchanged glances and, with considerable care, re-examined the control system. A mystery single wire from beneath the control system occupied their attention for two or three minutes. It led to a detonator stuck in the top of a case of Explosive 808 reposing in layer three of the pile.

It took five anxious minutes to make the huge pile safe, all of the explosive material now disconnected from the control system. They left the control system still switched on in case it was being monitored from a distance, but the chief interest now was what lay above their heads that was intended to be blown up. Like all Special Forces the two Frenchmen carried a Satellite Navigation System attached to their compass but, in the reinforced deep passage, it was not active. What they did know, however, was that the passage from the start ran due eastwards. All they needed to do was pace out the distance back to the compound shaft then head due east for the same distance on the surface.

They checked their watches and were surprised to find that they had been underground for almost four hours. They decided that the huge pile of explosives should remain guarded until it could be removed. The two French Commandos would remain until seven that evening, then Savage and Pierre would take the night shift. Food was not a problem because the kitchen back down the tunnel was well-stocked. Leaving their two French friends, Savage and Sarte returned the way they had come. It took almost an hour to reach the surface.

With Savage striding ahead and counting the paces and Pierre Sarte following in their vehicle making sure their direction was due east, the pair made a strange sight but, in twenty-five minutes, all was to be revealed.

Savage found himself at the boundary fence of the large fuel and gas storage facility, with about twenty paces to spare. Peter Savage and Pierre Sarte exchanged glances; what an explosion there would have been if the Gaddafi followers' plan had succeeded. Just three hundred yards away was a maternity hospital in the middle of a large residential area. All this would have been affected. They drove on to the Argus Hotel to bring their two female companions up to speed and get some rest before their shift was to start. Jane and Sally had spent the day with the Arabic translator and said that most of the paperwork referred to the urgent recruitment of mercenaries from as far south as South Africa. Savage produced another batch of paperwork he had found in the printer's tray that morning. Sally Gunn gave a sigh of exasperation,, as it meant another day with the translator.

Pierre Sarte went to the hotel manager and arranged an early meal for himself and Peter Savage to be served at five forty-five. Then they spent thirty minutes bringing the girls up to speed as to the contents in the passage lay-by and its intended target. They returned to the compound at a quarter to seven and made their way up the tunnel to where the two French Commandos stood guard. The senior of their new friends pointed out that at midnight, when the explosion was due to take place and nothing happened, would a group of Gaddafi loyalists come to see why?

Savage glanced at Pierre Sarte in annoyance; after all, with their experience they should have considered this. The two French Special Forces men had a solution to the problem. They would go back along the tunnel and prepare a meal in the kitchen, then rest in the bedroom there. At ten minutes to midnight they would return to the lay-by to make up a four-man reception committee. And so it was agreed and, a few minutes later, Peter Savage and Pierre Sarte were alone. An hour passed and the rich evening meal provided by the Argus Hotel was fermenting in Sarte's stomach. He released a great, rumbling but satisfying fart; it took just a few seconds to reach Savage and assail his nostrils. Savage almost choked as he complained: "Bloody hell, Pierre; that smell is awful and it would only take one electrical spark to ignite a fart and the entire pile of this fucking explosive."

The four or so hours before their friends return passed slowly; fortunately, Pierre Sarte's stomach at last began to settle down and the air became breathable once more. Just before midnight they heard their friends close by then, on the stroke of midnight, the control box gave a sharp click and the amber pulsing light reverted to red. The minutes following saw the four men on high alert, listening for any sound of returning Gaddafi loyalists. After an hour, they took it in turns to rest. This lasted until four in the morning, when there was the dull thud of an explosion further along the tunnel. A few minutes later a dense cloud of dust drifted down towards them, but then quickly settled.

At five in the morning Savage and one of the French Commandos made their way up the railway track to see if the tunnel was now completely blocked. At about one hundred and fifty yards they came across the site of the explosion. The tunnel was totally blocked and would require a considerable labour force to clear it if, of course, it was safe to do so. They returned to their friends and it was decided that two would stay while the others would arrange for the explosives to be removed. Pierre and the younger of the two Commandos made their way back down the tunnel to contact a Rebel Commander and arrange for the work party. Savage went to the kitchen area and made some coffee to accompany a large plate of buttered biscuits and some very mature cheese. He carried this back to their lay-by where he and his companion disposed of the entire plateful.

Just before eight o'clock, a great deal of noise and the sound of many voices coming up the passage suggested their work party had arrived. It consisted of about thirty men, who all expressed surprise when they saw the amount of material to be moved. They were, however, a cheerful bunch and set to with a great deal of shouting and laughter. It took three hours before the final box was removed, and a very dusty and tired Peter Savage and his companion were the last to leave. As they exited the massive, metal door, Savage relocked it but kept hold of the keys; they were to find a home in due course in the deepest part of Tripoli harbour.

Back at the Argus Hotel, both men went wearily to their rooms, showered and took to their beds until about four in the afternoon. The second batch of paperwork from the printer again referred to the recruitment of mercenaries, but with a larger pay package for loyal service. Jane was ordered to return to Benghazi, where she was to take charge of a security detail of British

Military Police in civilian clothing. Their role was to provide protection for a group of Whitehall civil servants on loan to assist the Transitional Government. Sally was sorry to see her go as they had become good friends but, on the plus side, she now had Peter Savage to herself.

Over the next few days many of the rebel army returned to their homes and businesses in Benghazi. Their replacements were far more professional and, though little was said at the time, many of them were in fact deserters from Gaddafi's regular army. At this time, the last major stronghold of Gaddafi supporters was Sirte. One or two smaller towns further south were the home of some tribal leaders whose loyalty was very much undecided so, for the time being, they were not being placed under threat of attack. The manager of the Argus Hotel had managed to repair his damaged top floor; the work had been carried out in record time. He had been quite happy to pay the inflated prices for the work because he could easily make up the cost with the increasing demand for his rooms.

The world's press and a large contingent of television presenters had now descended on Tripoli determined to keep the world aware of the dramatic events unfolding there. Peter Savage viewed the news industry with a cynical eye. His opinion was shared by most servicemen, whose experience suggested that the media never allowed the truth to stand in the way of a good story. Peter Savage remembered how, during his time in Northern Ireland, a journalist friend had responded to his editor's demand that he submit a story showing the current Conservative leader at the time in a good light. The journalist went fishing for three days and submitted absolutely nothing to his paper. His editor, realising his mistake, remained silent.

One of the American TV reporters from Fox News was anxious to find and interview Abdel Basset Al Megrahi, the only person ever to be convicted of the Lockerbie bombing. Having been told by a fellow journalist that Savage was a British Government agent, he approached him and Sally Gunn with his question as they went into dinner that evening. Savage viewed the overweight and perspiring reporter with obvious disgust, but just managed to say, quite politely: "I am sorry, but I cannot help you." The reporter was not used to being dismissed in this way and said angrily: "Just state your price, Mister; that should help loosen your fucking Limey tongue."

Savage reacted instinctively with a jolting left jab and a crisp right cross, sending the gasping reporter flat onto his back. Sally grabbed his arm, fearing that Savage intended to extend the lesson, but it was all over. Two of the hotel staff helped the dazed reporter to his feet and assisted the staggering figure to a chair. As Peter and Sally moved past him the reporter managed to gasp painfully: "I won't forget this, you bastard. I have friends in high places." Pierre Sarte, who had witnessed the complete sequence of events, leaned over the damaged reporter, his massive frame casting a huge shadow. "Please do not threaten my friend; he too has friends in high places and goes grouse-shooting with Prince Philip every year."

Later, as Sally, Peter and Pierre were leaving the dining room after a very tasty meal, they were approached by the hotel manager. He smiled and informed them that the unpleasantness they had been subjected to earlier that evening would not be repeated, as the Fox News reporter had been asked to leave and had done so. As they reached their landing, Pierre said: "Goodnight" and went into his room. At Sally's door Savage paused, and was about to suggest an intimate drink when a huge explosion rocked the hotel. Glass splintered and the pair, who had been thrown to the floor, became covered in dust and plaster fragments. Sally coughed and then said: "I could do with a drink, Peter, but I expect the bar is now closed."

Chapter Thirty-Five

Over the next few days several large arms-dumps were discovered by the ever-curious rebel fighters. After almost blowing up the Argus Hotel they had been more discreet in disposing of their dangerous haul, taking it out into the desert for detonation. Most of the world's press and the majority of the population of Libya were just waiting for the end to come for the Gaddafi era. The former leader still had support but it was ebbing fast and internationally he had been deserted. In a surprising development, a mysterious mound of signals dating back to the previous British Labour Government had been found littering the floors of the deserted British Embassy. The signals related to the then relationship with the Gaddafi administration and covered meetings and the handing-over of people hostile to Gaddafi.

The Tory press in the United Kingdom clearly had been given privileged access to this material and took full advantage in seeking to embarrass members of the previous administration. Many people, who knew how careful our embassies abroad were in disposing of any material likely to embarrass their Government, smelt a rat, and a very large one at that. Another matter being politically exploited was the future of Al Megrahi, generally thought to be in the last few weeks of his life. The British Prime Minister was on record as saying that Megrahi should never have been released from prison, knowing this statement would be favourably received by the Americans. The Scottish First Minister was quick to say that the release of Al Megrahi was a matter solely for the Scottish Assembly and no one else.

Al Megrahi was the only person convicted of the Lockerbie atrocity, which resulted in the death of 270 people in December 1988. Most of the deaths were the passengers of Pan Am Flight 103, which was blown-up over Lockerbie. The other deaths were people killed on the ground by the falling wreckage. Who was responsible for the blowing up of the plane and why has been, and still is, a matter of considerable controversy. In July 1988, during the Iran-Iraq war in which America and Great Britain supported Saddam Hussein's Iraq, American and other countries' warships patrolled the Persian Gulf. Their role there was to safeguard the passage of oil tankers to and from the Kuwait oil terminals.

One of the warships patrolling the Gulf at the time was the USS Vincennes, a destroyer carrying surface-to-air missiles. It was in company with two other destroyers, one American and the other Italian. On this particular July day, all three warships were alerted by their radar operators that an aircraft had been detected leaving the Iranian coast. This aircraft was quickly identified by two of the warships as a passenger plane, probably an Airbus on a scheduled flight from Iran to Mecca in Saudi Arabia. The Captain of the Vincennes did not agree with their decision and decided that his command was under threat from this aircraft, and ordered two missiles to be fired at the intruder.

Both missiles struck the aircraft and the Airbus crashed into the Persian Gulf, killing all 290 passengers and crew. Among the passengers were thirty children. Most of those killed were on their way to Mecca to fulfil the lifetime requirements of the Koran. In the enquiry that followed the Captain of the Vincennes, Captain Will Rogers, insisted the aircraft posed a threat to his command and that he was entitled to respond. The Captains of both the other ships insisted the passenger aircraft was easily identifiable and the Captain of the other American warship went further by saying that Captain Rogers had a reputation for being overly aggressive.

Five months later when the Lockerbie incident occurred, most of the world's press, for one reason or another, chose not to connect the two tragedies, nor did the British and American Governments. Instead, investigations carried out by the two countries suggested agents from Libya as being responsible. Key evidence, recovered from the wreckage in very close proximity to the explosion on the aircraft, was some cloth thought to be in the same suitcase as the explosive device. Subsequent investigations suggested the material had been purchased in Malta, but by whom? A painstaking search of shops selling this material was centred on one particular shop in Valletta. Initially the proprietor said he had no recollection of selling the cloth, which was part of his stock, so he could not identify the buyer.

Libya had at the time invested quite a lot of money in the hotel industry in Malta, and Libyan businessmen were welcome on the island. Libyan Government officials and their security officers were also frequent visitors; among them was a certain Abdel Basset Al Megrahi. This gentleman was well-known to the CIA and MI6 and was considered to be a top Libyan agent. At this point it would appear that a photograph of Megrahi was shown to the shopkeeper by the American investigators, and it was suggested that this person was the likely purchaser of the cloth. The shopkeeper again said he did not know who had purchased the cloth or recognise the person whose photograph he had now been shown.

It would appear that at this point it became clear that in America, certain people were prepared to pay a very substantial sum of money to any witness who could identify anyone involved in the destruction of Pan Am Flight 103 over Lockerbie. The shopkeeper at this point, after a great deal of thought, suddenly remembered who had purchased the cloth. It was the man whose picture he had been shown. The rest is history, with Al Megrahi being found guilty of involvement in the destruction of the aircraft. Another factor to cause concern over Megrahi's conviction was that the vitally important witness had now disappeared from Malta and was thought to be in Australia, now with a completely new identity.

Another matter in the Megrahi saga is worth a mention. When his health went into decline he was about to have an appeal heard against his conviction. Many people who had taken an interest in the case considered he had an excellent chance of winning that appeal. But when his medical team produced their medical evidence that his life was almost at an end and that he should be released on compassionate grounds, the courts insisted on a condition for his release. This condition was that he should abandon his appeal against his conviction, or he would not be released on compassionate grounds. Now, with Colonel Gaddafi, his protector, being forced to flee the rebel advance, Megrahi's freedom is once again under threat, despite his life being close to its end.

Sally Gunn and Peter Savage had been instructed to make contact with any former Gaddafi supporter who could provide information about secret arms-dumps and underground control-rooms. The international oil companies had expressed concern that their pipelines and pumping stations would be a final target for Gaddafi's revenge. One morning, as Savage and Sally Gunn were finishing breakfast at the Argus Hotel, the waiter told them that a rather scruffy individual wanted to see them on the hotel steps. The waiter added that the manager would not allow this person to come into the hotel.

Finishing their coffee the pair, now curious, went outside. There, sitting on the bottom step, was their visitor. The man was old and dressed in a torn and filthy desert robe. Savage led the man to a garden seat shaded by a mature Cypress tree. All three sat down, Sally keeping Savage between herself and their visitor. The old man took a deep breath and spoke: "Gaddafi and his murderers killed my only, much-loved son. I know where an underground room of great importance is, and taking you to it will be my revenge." Savage turned to Sally. "I think we will involve Pierre in this; can you go and find him?" Sally made her way back into the hotel as Savage questioned the old man further as to the location of this secret room.

An hour later, with Pierre Sarte driving and all the windows of the vehicle open in an effort to dilute the old man's powerful smell, they headed south from Tripoli. With the outskirts of the city still in view, their guide indicated they should stop at a ruined farm building. They parked their vehicle out of sight from the road and followed the man into the rear of the old building. In one now-roofless room he indicated they should move some rubble-covered doors. They quickly did this, to disclose a square of new, green canvas, which the old man pulled aside. There before them was a fairly new trap-door about four feet square. "Down there," said the old man. Pierre glanced at Peter Savage, who nodded, and both men lifted the door open on its hinges.

With the trap-door now fully open, a set of concrete steps could be seen, leading down some twenty feet or so into complete darkness. Sally had brought a powerful torch from the car and led the way down into what lay below. Savage was immediately behind Sally, followed by their guide and, lastly, Pierre Sarte. At the bottom of the steps was a concrete-lined tunnel and there, ten paces further on, it was blocked by a metal door similar to the one they had encountered beneath Gaddafi's compound. This one, though, had a different locking system and they were confronted by a digital dial with two light sources, one red, which was illuminated, the other obviously green, which would illuminate should the correct combination be entered.

Their guide moved forward and, without hesitation, punched in four numbers. The red light went out and the green light flashed on. Savage and Pierre Sarte went to put their weight to the door but they were stopped by the old man, who said: "There is a delay of thirty seconds; then it will open." Savage counted up to twenty-five; then there was a hiss and the door swung effortlessly open. An illuminated passage lay before them, with the concrete floor sloping down to another door. At this door their guide paused and said: "As we turn this door handle the outer door will close but, when we exit on our return, the outer one will re-open."

The second door opened as they turned the handle and their guide was correct, as the outer door shut. The room they now entered contained banks of computers, with multi-coloured lights flashing everywhere. The old man moved forward and pointed to several groups of lights and explained that, if they were now showing red, a charge had been activated at some oil terminal or pipeline junction and would explode within the hour, and there was nothing they could do to prevent this. However, if the light was still green they could switch it to safe and prevent it exploding. Sally moved forward, her nimble fingers racing over the keyboard and turning many of the green lights to amber. As she worked, some of the other indicators showing green turned red, completing Gaddafi's revenge on his former friends in the oil industry.

Chapter Thirty-Six

The three friends and their foul-smelling helper and guide spent another hour in the hidden control room. "There is money here as well," said their guide "I wish to take it to support my family and my son's wife and children." Savage looked at Pierre Sarte, who gave a dismissive shrug. "Where is it?" enquired Peter Savage. The old man led them into a small room at the back of the computer room and pointed to a wall cupboard about two feet square, but with no obvious locking device. "It could be booby-trapped," offered Sally helpfully, as she moved back to the doorway. Savage moved forward and examined the two doors; each one had a small, brass door-knob. Taking a piece of notepaper from his pocket, Savage folded it into a narrow strip and carefully inserted it in the tiny gap between the doors. He then slid it down the gap from top to bottom, and then repeated the manoeuvre to the sides and the top and bottom of the cupboard.

Sally gave an audible sigh of relief when he had finished. Savage then pulled the doors open. On one side of the cupboard was a medical box clearly marked by a red cross and, on the other, was a pile of treasury notes, mostly Euros. Savage took the notes and handed them to Sally to check and sort out the various note values. The majority of the currency was in one-hundred-euro notes and about a quarter in notes of fifty Euros. The grand total was two hundred and fifty thousand Euros. Nothing was said as Sally Gunn put the prize in a canvas bag and handed it to the old man. They made their way out, the outer door opening as they closed the inner door. Both Savage and Pierre Sarte had thought of destroying the command centre, but it had to be revealed to the rebel fighters in due course.

They drove back to Tripoli, dropping-off their guide near the ruined police station then drove on to the Argus Hotel, hoping to be given a late lunch. There was some excitement in the hotel, and the television sets were surrounded by eager watchers as pictures were shown of several oil-pumping stations ablaze and news of several pipeline explosions. The three friends exchanged glances and almost guiltily went into the dining room. Savage, in his gentlemanly way, assisted Sally into her chair, the view of her neat and tightly-tailored backside being his immediate reward. To celebrate their successful morning they treated themselves to some ice-cold orange-juice to accompany their lunch. Their task later on that afternoon was to pass on the location of the control room to their rebel friends who, no doubt, would use some of their vast supply of captured explosive to destroy it.

A British warship had been given permission to enter Tripoli harbour. It was carrying a large amount of medical supplies donated by the people of Malta. Peter Savage took the opportunity to visit the ship and speak to the Captain, hoping to talk to London. The Captain was only too pleased to help, and soon Savage was briefing MI6 on the current situation in Tripoli. He also received an update on his role in Libya now that the rebels were in command. He was surprised to hear that the capture of Colonel Gaddafi was no longer at the top of the British wish-list. His death would be considered a satisfactory outcome but, alive and free to speak at an international court, there would be a great deal he could say that would embarrass the British Government.

Another matter that was raised was the use of memory sticks that could be set to alter a previously-set computer program. Savage was reminded that the Chinese were considered world leaders in this devious practice, with an ultimate aim of being able to monitor any country's nuclear weapon plan. Savage gathered he was likely to be moved to the Far East shortly, though no date to leave Libya had been suggested. Savage made his way back to the Argus Hotel, hoping to spend some time with Sally and make some progress in the physical sense. He arrived just in time to say farewell to Pierre Sarte, who was to go back to Benghazi and resume his love affair with the lovely Suki Hodges.

That evening at the Argus Hotel an after-dinner dance had been arranged for the guests' enjoyment. Most of the females resident there were from the television companies, able now to look pretty good as they told the viewers of the horrors confronting them in Tripoli. As the evening wore on, Savage danced with Sally several times also allowing her to dance with other male residents. Sally Gunn was becoming frustrated with the too-gentlemanly Peter Savage, hoping that at least he might attempt to touch her breasts. Just before midnight, as they danced together, Sally thrust her hips forward, making full contact with the lower part of the Savage body. Instantly, Sally felt his very evident interest, and quite a lot of it. Nothing was said, but both of them knew how the evening would end.

At twenty minutes to one o'clock they were in Sally's room, both shedding their clothes in the dim, night light. Savage leaned forwards, his lips caressing each of her breasts in turn. Sally grasped Savage's rigid shaft, pressing it against the top of her legs. Once on the bed, Savage wasted little time in inserting himself fully into Sally's eager body. After a frantic few minutes of activity, Savage suddenly paused; then Sally felt the result of their lovemaking pour into her eager body. A strange shyness overcame them both, but it did not last. Sally's exploring hand discovered a stiffness returning to the Savage member. It was four in the morning when Savage left Sally's room with a final farewell kiss. Both knew that now their relationship had changed forever, and both were comfortable with the new arrangement.

Savage was first down for breakfast and, when Sally joined him ten minutes later, he stood up and greeted her with a kiss. The moment was not lost on Sally; she had given herself to this man, and now his kiss made it all worthwhile. Two days later the news both had been dreading arrived via the temporary British Consul. Sally was to return to London and Savage was to go back to

Pakistan and then India. They made the most of their last few hours together. Both travelled from Benghazi to Rome, then making their separate ways to London and Karachi. Anxious to tell Pierre Sarte of their new relationship, they found no trace of him or Suki Hodges in Benghazi.

On arriving at Karachi, Savage was met by a very smart young police Lieutenant, who treated him with great respect, taking him to one of the city's best hotels. He informed the pretty receptionist there that Major Savage was an honoured guest in Pakistan and should be refused nothing. The girl looked at Savage and fluttered her eyelashes, saying to the police Lieutenant: "I completely understand, Sir." For just a moment the new relationship with Sally Gunn was in jeopardy, but the moment passed, leaving Savage feeling very proud of his new self.

The next morning the smart police Lieutenant collected Peter Savage from the hotel at ten o'clock, enquiring if he was completely satisfied with the service. Savage said he was, thanking the Lieutenant for his concern. He was delivered to a very smart and obviously newly-constructed building which was well-guarded by police and military. An ultra-fast lift delivered him to the top floor and he was shown into a room already half-full of a mix of civilian, military and police, all involved in the intelligence business. Peter Savage knew several of those attending by sight, his presence being acknowledged by a raised hand or a nod. Coffee was served and the Chairman took his seat, using his silver cigarette-case to call the room to order.

The Chairman remained seated as he spoke. "Good morning, my friends. It is now three months since we last met, and much has changed in the world in which we operate." He then went on to speak about the 'Arab Spring' and the regime change in Libya, unrest in Syria and the never-ending problems between Israel and the Palestinian people. "I have some good friends in Turkey," said the Chairman. "They are not prepared to put up with the Israeli aggression in preventing food and medical aid entering Gaza; in fact, I am told the next supply ships to Gaza will have Turkish warships to guard them." Iran was the next country to be mentioned, with information that the nuclear facility there was now producing electricity for the national supply.

The Chairman spoke for an hour, then there was a short break and they reconvened with everyone who wished to given the opportunity to speak. They then took lunch but, when they went back to the room, the police had returned to other duties. Savage was surprised to be singled out by the Chairman in his first comment of the afternoon, when he said: "Tell me, Peter; you are going to India from here; how much of today's business will be passed on to our neighbours?" Savage smiled and replied: "My country considers Pakistan and India as good and trusted friends and, in my small way, I am tasked to keep it that way; not easy as we just beaten them at cricket." A ripple of laughter went around the room and the Chairman smiled and said: "But you did not answer my question."

The meeting finished just before four o'clock. Savage managed a few words with the Chairman before he left, asking if there was anything he could pass on to the

Indian intelligence service when he arrived in Delhi the next afternoon. The Chairman smiled and shook his head. "No need, my old friend; they will have a copy of what we discussed today by then. We mutually leak - it reduces the tension." Savage was escorted back to his hotel by a female Inspector of Police, who said she could show him the best nightlife in Karachi that evening if he was interested; but he declined. As he collected his room key the attractive receptionist smiled and fluttered her eyelashes; the gesture was full of eastern promise.

That evening after dinner, Savage returned to his room and watched some television. Last of the Summer Wine was being shown with Compo speaking in Hindi; Savage found it hilarious. He slept well and, with his flight to Delhi scheduled for midday, took his breakfast about eight o'clock. When he went to book out the attractive receptionist was not there; her replacement was a very slim young man. He fluttered his eyelashes at Savage and assured him his account had been settled and that a driver was waiting to take him to the airport. At the airport, Savage was spared the normal hassle of booking-in and was taken straight to the VIP lounge. He was given coffee and driven out to the waiting 737 before the other passengers had boarded.

They took off on time for the seven-hundred-mile flight, which should take about ninety minutes. Ten minutes into the flight and with clear skies, Savage was able to look down onto the Thar Desert, which in the past was home to most of the caravan routes. Looking round at his fellow passengers, most were young student types who were about to change their place of learning. The plane was only half full and the two stewardesses had time to smile and talk to their passengers. Twenty minutes later Savage noticed a change to the sound of the plane's engines; then the starboard engine stopped altogether and the seat-belt sign came on. One of the stewardesses hurried forward to the flight deck, but emerged seconds later looking far from calm.

As she and her fellow stewardess checked that their passengers all had their seat-belts secured, the other engine stopped. The plane lurched to the left then the pilot, anxious to keep his airspeed high enough to have some control of the aircraft, tilted the nose of the 737 downwards. The pilot's voice came over the loudspeaker, quietly informing the passengers to brace themselves and to remain seated until the plane came to a halt. Savage glanced out of his window.

They were now at about five hundred feet and the Thar Desert below them was a mass of small humps. The nose of the aircraft lifted slightly and the flaps were now fully deployed in a last desperate effort to reduce their speed. Contact was made with the ground in a screech of metal and a bounce that lifted them all in their seats. They grounded again with an appalling crunch, then the aircraft spun as one wing contacted a desert hump, but they slid noisily onwards. A sudden, mind-numbing jolt brought the plane to a standstill and, in the momentary silence that followed, they became aware of the overpowering stench of jet fuel.

Chapter Thirty-Seven

One of the stewardesses acted promptly, opening the rear door on the side where a wing was still in place. Her companion, who was responsible for the forward door, lay unconscious in the gangway. With the rear escape-chute deployed, there was an indisciplined rush by the passengers down the length of the plane, the unconscious stewardess being trampled in the process. Savage went to the aid of the fallen stewardess; as he did so, he was puzzled that the two flight-deck crew had not appeared. He dragged the still-unconscious flight attendant to the aircraft's rear door and placed her feet-first at the top of the chute. The other stewardess had shepherded the other passengers to a place of safety some hundred yards from the crashed aircraft.

In response to Savage calling to her, she came across with two young men to take delivery of her colleague. With a reception party now in place, Savage eased the stewardess onto the chute then released her. The still-unconscious girl slid down, giving the appreciative young helpers a two-second glimpse of her powder-blue panties. At the chute bottom her colleague quickly restored her friend's modesty by adjusting her skirt. Savage made his way to the flight deck, to find the door was ajar but jammed. Using both hands and his considerable strength, Savage managed to force the door fully open. Both pilots were slumped forward over the controls with their seat-belts still in place.

In his career Peter Savage had encountered death many times, and it took just a brief examination to see that the senior pilot was dead. As Savage felt for a pulse in the second pilot, the man began to mutter and attempt to release his seat harness. Peter Savage helped him to stand up then assisted him the length of the aircraft to the rear door. As Savage tried to ease the man onto the chute the second pilot, whose face was bleeding freely, resisted; then he whispered: "Crew last, my friend, passengers first." Savage slid down the chute and waited for the injured second pilot to join him.

Back at Karachi in the control-tower complex, the air traffic controller was about to hand over the Boeing 737 to Delhi when the pilot had reported the loss of one engine and then, a few moments later, the remaining one. The controller immediately alerted the Emergency Services with the plane's approximate position. Because of the area over which the plane had been flying, two helicopters were dispatched, one a Chinook with a fire detail, the other slightly smaller with a medical team aboard. As Savage and the injured pilot reached the passengers, the plane's escaping fuel vapours encountered a heat source. There was an ear-shattering 'crump', and a column of flames and smoke rose into the still, desert air.

When this occurred the two helicopters were just a few miles away and immediately saw the smoke from the Boeing's funeral pyre. The varying nationalities in the two rescue teams quietly called on their various Gods to keep the passengers and crew safe. A few minutes later the two helicopters landed and their rescue crews rushed to the passengers' aid and were delighted to find them all, bar one, safe and well-clear of the burning aircraft. Just two of the crew required medical treatment, and none of the passengers. The second pilot had his facial cuts attended to and his very sore, bruised chest strapped up. The stewardess was suffering from concussion and a badly bruised back, probably caused by the escaping passengers. She was now conscious and was being comforted by the second pilot, who was her boyfriend.

An hour later and with the fire out, a considerable achievement considering the limited powdered chemicals available on the Chinook, it was now time to take the passengers back to Karachi. With the light failing, ten passengers would have to spend the night at the crash site, as the Chinook and the other helicopter could not take them all. Savage was quite happy to stay and the uninjured stewardess considered it was her duty. Eight other volunteers were quickly found, all anxious to prolong the adventure. They were given food and water, a canvas shelter and top-of-the-range sleeping-bags. The remaining ten watched as the two helicopters flew off in the failing light, leaving them to prepare a meal and shelter from the quickly cooling Thar Desert air.

With the canvas shelter erected, a meal was prepared by two of the young passengers, who were studying catering. An hour later, with a satisfying meal inside them, they all settled down for the night, drinking a final mug of tea or coffee. The sleeping-bags kept out the chill desert air and most of the survivors managed a reasonable night's sleep. Breakfast consisted of tea or coffee and a large cereal survival bar. As it was being consumed they heard the sound of the approaching Chinook helicopter.

As the helicopter landed there was no rush for a seat inside, as they all were aware that the remains of the Senior Pilot needed to be recovered and accompany them back to Karachi. They packed up the canvas shelter and the remaining survival equipment, and then stood silently as the pilot's remains were placed aboard the Chinook. Ten minutes later they were on their way back to Karachi, with one last look at the burnt wreckage of their aircraft. No doubt in the next few days an investigation would take place to explain why both engines had become starved of fuel when the Boeing's tanks were full.

After a noisy but uneventful flight in the Chinook they landed at Karachi. Savage was surprised to find the very smart police Lieutenant in attendance on him once more. Giving Savage a very precise salute, he told Savage that he was booked in to the same hotel. He also said that Major Savage was required to fly to Lahore instead of Delhi the next day. Once back in his hotel room, having been given the fluttered eyelash treatment once more, Savage worked out what kit he needed to replace that lost on the Boeing 737. A phone call interrupted his thoughts - it was a very concerned Sally Gunn. Having satisfied her that he was fine, she passed on instructions from their boss that he was to visit the Lahore Atomic

Research Centre. He was expected and all would be explained there.

After a late lunch Savage took his list of clothing replacements to the local branch of Brooks Brothers, paying by his credit card. He selected three shirts to replace the two lost, and did the same with his underclothes. Shoes, socks and a lightweight, grey suit completed his clothing purchases, and then a very smart, lightweight travelling case was added to the bill. "We will deliver them to your hotel, Major Savage," said the young salesman "It has been a pleasure to serve you." "You know where I am staying, then?" said Savage. "Of course, Sir," said the young man "We were told to expect you." At four o'clock that afternoon his purchases were delivered to his room by one of the hotel staff.

When Savage went down to dinner that evening, he was shown to a table occupied by a man who had his back to him. As Savage reached the table the man stood up, a smile lighting up his face, the right side of which was badly scarred. Savage grabbed the man's hand, his delight obvious. "My dear Asif; it must be five years since we last met. What a very nice surprise." The two men were in deep conversation throughout the meal. Their friendship had begun when Savage was serving in Afghanistan in his last tour in the Royal Marines. When they reached the coffee stage in the meal, Asif told Savage he would join him on the trip to Lahore, saying: "They have a problem there, my friend; we are hoping you can help solve it."

The two men parted at eleven that night. Asif Ahmand told him he would see him at the airport the next morning. Savage took to his bed a short time later, wondering what the problem was at the Atomic Research Centre. Savage woke once during the night as a tropical storm passed over the city. He awoke at seven and went down to breakfast just before eight o'clock. There were only six people taking breakfast, but the air was full of the delicious smell of coffee. At nine-thirty Savage was down at the reception desk, where the smart police Lieutenant assured him his account had been settled. He was then driven to the airport, where he found Asif Ahmand waiting for him in the reserved part of the departure lounge.

At ten forty-five their plane took off. It was another Boeing 737, but in the ninety-minute flight the engines never missed a beat. A car was waiting for them at the airport arrivals section and they were driven straight to the Atomic Research Centre. They were greeted by the Deputy Director, who first took them to lunch in the senior managers' canteen. As there were quite a few people present, no mention of the problem was made; this was not raised until, having completed their lunch, they were in the Deputy Director's office. "We appear to have a bug in our computer system, Major Savage. The programs we wish to run are mysteriously amended after we have set them in place."

Savage recalled the intelligence he had seen of the Chinese operation in producing 'memory sticks' specifically programmed to do just that. He was careful not to jump in with both feet, and innocently asked how frequent were the security checks on the people having access to the main computer. The Deputy Director looked slightly ill-at-ease and confessed this was left to his Centre's Security Chief. "Can you find out now?" asked Savage, aware that Asif Ahmand was furious at the Deputy Director's laxness. The Deputy made a phone call asking the question. The answer was fairly short and he replaced the phone. Before telling Savage and Asif Ahmand what he had been told, he wiped his forehead and then blurted out: "Once every year."

Asif was not to be ignored and told the Deputy it should be standard practice to carry out checks every three months, according to the level of secrecy involved. "And how often are your key personnel's bank accounts checked?" enquired Savage, his voice quiet but full of menace. The now-pale-faced Deputy made another phone call and, on receiving an answer, slammed down the phone, his hands shaking in anger. "They say that they do not check key personnel's bank accounts unless they have written permission to do so from the account-holder." Savage burst out laughing then said: "I suggest your Security Officer gets off his arse and does some checking without the holder's permission; being bloody devious is all part of his job."

Asif and Peter Savage spent another hour with the Deputy and then they were driven to their hotel. They had been placed in the VIP annexe, a single-storey building in a quiet part of the extensive gardens. It had a security guard patrol twenty-four hours a day, which reflected the importance of its visitors. They had made arrangements to return to the Centre the next morning; in the meantime, Asif had obtained the police records of all the staff with a high security rating.

Chapter Thirty-Eight

In the three hours before dinner, Asif and Savage went through all the security files of the top staff at the Centre. They were not surprised to find little in them to raise their suspicions though two of the staff, both women, holidayed regularly together in Hong Kong. Asif and Savage looked at the photos of both women, neither of whom was married. One had a face like a sour apple, the other was extremely attractive. A strange pair to be friends, and worthy of a more selective check. The residents in the annexe used the hotel's main dining-room, though their dining area was discreetly roped off. The food was excellent, as was the service. Their waiter was a young, good-looking man; Savage noticed that he waggled his arse more than the waitresses did. He mentioned this to Asif who said he had not noticed, which Savage thought was strange as his companion's eyes had followed the young waiter everywhere.

After their meal they took their coffee in the annexe and discussed their next visit to the Atomic Research

Centre. A thought suddenly struck Savage and he asked Asif where his country's nuclear weapons were stored. Asif laughed and said he did not know, but thought they were moved to different sites on a regular basis. He then asked Savage where his country's nuclear stock was housed. Savage admitted that, apart from a submarine being constantly on patrol armed with nuclear missiles, he had no idea where the rest of their missiles and bombs were kept.

After breakfast the next day they were driven to the Research Centre, where a very tired Deputy Director, who clearly had spent the night in his office, said the various bank accounts of his top-rated staff were now available for scrutiny. The three of them spent the next hour running through them, finding not one trace of any suspicious movement of money. One thing that did surprise Savage, but not Asif, was that the pay of women in the same grade was lower than that of their male counterparts. They were then taken to the main computer room, where a great deal of the secret research was initiated and stored. It was pointed out to both men that any memory stick inserted into the system was reported by entry into the day's log and recorded on the twenty-four-hour surveillance system. Dr Khalid, who was the senior member of the computer team, tended to dismiss the theory of a memory stick deliberately introducing a disruptive program. He had an old-fashioned view that, if something could go wrong in any man-made object, then at times it will.

During the morning Asif and Savage had the opportunity to see most of the staff with access to the computer. Both women who holidayed in Hong Kong were there, but did not seem very much like their photographs. The one thought to have a 'sour apple' face was, in fact, quite pleasant-looking, and her friend turned out to be not quite as attractive as at first thought. While they were in the computer room there was a malfunction in one of the computer's program which, if it had not be spotted and corrected, would have released some radioactive water into the normal waste water system.

After lunch both men looked through the security films of the computer room covering a twenty-four-hour period before a problem had occurred. It was a boring task and it produced not the slightest clue in the probability of a rogue memory stick being introduced into the system. The two men returned to their hotel just after four in the afternoon. The hip-swinging waiter was placing out clean table linen as they passed by, and he and Asif exchanged smiles. As they reached their rooms, Savage could not contain his curiosity and asked Asif a straight question. "Do you find the waiter attractive, my friend?" Asif looked surprised then answered: "I find most young men and women attractive, and I do enjoy intimacy with them both." Then, seeing the surprise on Savage's face, added: "But do not worry, Peter; you are too old and tough-looking for my taste."

That evening after dinner the guests were entertained by two singers, who performed songs from the musicals singly or as a duet. The woman, who was in her fifties, clearly had been very attractive in her younger days. Her companion turned out to be her son. It was just after ten-thirty when the concert finished; it had been quite excellent and the applause reflected that. The young waiter came over to their table and asked if there was anything they required. Asif was about to reply when a great blast of energy lifted them and the room's tables into the air, and they were deafened by a huge explosion.

Savage, who had been hurled back against a dividing wall, struggled to gather his senses. He attempted to move his legs but they seemed to be fastened to the floor. Flames were devouring the curtains that were draped around the walls and as Savage began to regain his hearing he could distinguish the agonised cries for help. He resolved the problem with his legs - they were trapped under an occupied chair. After a struggle, Savage slid out from under the weighty object, whose occupant was Asif Ahmand, who was covered in blood. Savage was relieved to see that Asif was breathing but still unconscious. Another sound reached his ears - it was the comforting sirens of the emergency response vehicles.

Within minutes firemen were spraying water on the still-burning debris of furniture and helping the injured from the room. With another survivor's help, Savage managed to get a dazed Asif from the dining room and out through the totally wrecked reception area of the hotel, onto the road outside. Bodies lay everywhere and Savage began to realise the magnitude of the explosion, whose point of detonation appeared to be in the hotel's reception area. Like its neighbours to the east and west, Pakistan was no stranger to this type of incident. The suicide bomber, usually a Taliban volunteer, operated in Afghanistan and India, his objective to cause chaos and destabilise the existing authority.

Asif Ahmand was now conscious but did not respond to questions regarding his injuries, which did not appear serious. Savage and a medical orderly managed to get him into a police van that was taking the less-seriously injured to one of the city's hospitals. Savage wondered if the annexe had escaped serious damage, and made his way back through the hotel to his accommodation. Apart from the entrance door being blown open the rooms, though without power, were almost undamaged. Feeling somewhat guilty over his concern for his own comfort, Savage went back to the dining room and assisted in the removal of those who had lost their lives. He helped a young police constable carry out several bodies, which they wrapped in blankets to conceal their appalling injuries. Sadly, one of the dead was the young waiter Asif was fond of; he was missing a good deal of his left leg but his handsome face was untouched and his eyes were still wide open, as though waiting for a question to be answered.

The hotel's manager spoke to Peter Savage, offering to transfer him to another hotel, but Savage had decided to stay. The hotel's kitchens were untouched as they were below ground, so the preparation of food was not a problem. It was just after two o'clock when Savage managed to get to bed, and he fell asleep almost straight away. He awoke at seven and, after a cold shower and a shave, made his way to the dining room to enjoy what they had managed to muster for breakfast. The remaining hotel staff had replaced all the damaged furniture and, though there were fewer tables, all were covered with pristine table linen.

After breakfast Savage decided to find out where Asif Ahmand had been taken. To his surprise a list was pinned to the temporary reception desk showing the location of all the injured hotel guests. Asif was being treated in the local maternity hospital, which had opened a spare ward to assist in the emergency. It was only a ten-minute walk from the hotel and Savage was glad of the exercise. A stern-faced nurse at the hospital's entrance pointed Savage in the right direction and within minutes he was standing next to the dozing Asif, whose face was a mass of small cuts covered in iodine.

"How are you my friend?" enquired Savage. Asif's eyes opened and he gave what passed for a smile. "Your voice seems very soft and almost from a distance, but my hearing is improving," said Asif, struggling to sit up in his bed. The two friends talked for some time, mostly about the problem at the Research Centre. Savage made no mention of Asif's waiter friend and Asif never asked. As they were talking, a doctor visited his unusual patients. He gave Asif the once-over and said: "Well, I think you can go home; your hearing is undamaged, just temporarily dulled, and will improve over the next few days." And as he turned to go he added with a smile: "And I can assure you that you are definitely not with child."

Savage arranged for a taxi and, after they had thanked the matron at the hospital for her staff's kindness and care, the taxi took them both back to the hotel. They both spent most of the day catching up on their sleep. Savage telephoned the Research Centre in the late afternoon explaining their absence, and explained they had been caught up in the hotel bomb-blast. The Deputy Director told him that both the ladies who holidayed in Hong Kong had not reported for duty, adding that this was most unusual as they were both exceptional timekeepers. Savage phoned his London office and spoke to Sally; she knew of the hotel explosion and her voice betrayed her concern. After replacing the phone he decided he would spend one more day in Lahore then take the plane back to England.

When Asif and Savage met for dinner that evening they found the dining room almost back to normal. The main course was fish, and their attentive waiter boned the dish for them at the table. It was at this point that Asif asked the young waiter if their regular waiter was off-duty. The young man finished his task with their fish then, with tears running freely down his face, told of his colleague's fate. The meal was completed in silence and, after declining coffee, Asif said an abrupt "Goodnight" and went to his room.

At breakfast the next morning nothing was mentioned of the previous evening and, just after nine o'clock, they were collected by car and taken to the Research Centre. There the Deputy Director greeted them with the news that the two women employees had flown the previous evening to Mumbai. With no official contact between Pakistan and India, the two women could not be returned and may have already moved on, perhaps even to China. The Research Centre was already reviewing all of the women's input into the main computer and instigating a search of their rented flats. Savage thought this should solve the Centre's problems; where radioactive materials were concerned you could not be blasé about security.

Later that afternoon he said "Goodbye" to Asif Ahmand and made his way to the airport, looking forward to seeing Sally. The flight was only three-quarters full and he was upgraded to Business Class without even asking. The flight was uneventful and on arriving Savage, who had not adjusted his watch, found that locally it was an hour fast. He went first to his official accommodation and deposited his case, then up to his office to see Sally. With Madge his secretary on leave for a week, Sally was taking care of everything, and they managed a passionate embrace. They were interrupted by the telephone. It was the boss. Captain Ralph Maitland wasted few words: "I need to see you now. We have a meeting to attend in the morning and I want to be up to speed on your findings."

Chapter Thirty-Nine

In a remote part of Afghanistan just ten miles from the Pakistani border, twenty Taliban fighters were resting in a cave, deceptively large considering its small entrance. Several of the fighters had wounds that were being treated by the group's two female helpers. The women were not young and were the mothers of warriors in another group. The wounds were not serious and should heal quickly if they received even the most basic treatment. Anything more serious and the wounded man would accept his fate and cover his comrades' withdrawal by selling his life dearly.

The group's leader had trained as a teacher before committing himself to the Taliban cause. Their present task was to escort a young and very talented computer engineer to Kabul where, with Allah's blessing, he would sabotage the city's main computer. The young engineer had arrived three hours ago and was being allowed to rest because, when darkness fell, their journey would begin. A goat-meat stew was bubbling away over a small fire, its smell causing the warriors to give appreciative sniffs of the meal to come.

Only one moment had caused concern that day as, shortly after the young engineer had arrived, an American drone aircraft had flown up the narrow gully, its rotating eye seeking its prey. Two hours later and with their stomachs full, the men prepared to depart on their mission; most of the journey, taking three days, would be made at night. The only part to be completed in daylight was from certain villages where a lorry loaded with farm produce would carry them as part of its load. Not all of the twenty men would make the journey; ten were to lay an ambush at a road junction. The wounded men would be part of that group.

After four hours into their journey they rested for ten minutes, listening for any unusual sound; they then moved on. As they became aware of the first faint light of dawn, they looked for a place to rest up in the daylight hours. A dry river-bed, sparsely lined with some withered greenery, would suffice as long as they remained quite still. Two men went forward several hundred yards to warn of any movement ahead; if necessary they would sacrifice themselves to enable their fellow fighters to escape. The day was not too hot as the colder season was on its way and the nights were becoming much cooler. As the daylight began to fade the men prepared to move, hoping that by dawn they would reach a small village and fill their now-empty stomachs.

On day three the group was now reduced to ten as the ambush party had been left at the track junction, where they would lie in wait for three days. During this time their food would be supplied by a nearby village. At the end of the night's trek they were now at the village, from where a lorry would take them to the outskirts of Kabul. At that point only the leader and the young engineer would go on, as the road had several security checks to negotiate. The lorry driver was nervous but, with the leader sitting by his side posing as his brother, the journey began. Another matter very much in the driver's mind was the threat that, if he did not cooperate, his young wife would be taken into the hills to provide comfort for the Taliban fighters.

During the drive into Kabul they were stopped three times at check-points, two manned by the police and one by the military. At none of the check-points were they thoroughly searched, just a quick glance into the lorry's cab and a slightly longer look at the load in the back. At the rear of the market place the young engineer eased himself clear from under a pile of animal fodder and, with a brief word to the Taliban leader, began his walk to the computer centre. After a brisk, fifteen-minute walk he arrived at the entrance, which was guarded by both police and Afghan soldiers. A brief glance at his pass and a word with one of the soldiers, with who he had been at school, was enough to allow him to gain entry to this extremely important establishment.

A few minutes later, with a forged worksheet as his pass, he was in the main computer room, his small tool-case and a confident smile leaving him unchallenged. For a few minutes he checked various power sources, all innocent routine tasks for the engineer; then he moved over to the main computer. To avert suspicion he wrote down the computer's model number in his note-book then extracted the all-important memory stick from an inside pocket of his overall. In one unhurried and well-rehearsed move he inserted the programmed stick into the receiving slot and activated the transfer of its information. With the 'Trojan Horse' inserted, his task was complete.

The young engineer shut his small tool-case and, with a smile and a wave, left the building, the busy programmers not bothering to respond. A few hours later the inserted program began to take effect. Power surges caused havoc with nearly all of the many homes' electrical appliances. Fresh water from the main supply was reversed through the city's sewage system. But with people desperate to report their plight, all of the many communication channels were in chaos; all that could be heard was a very loud 'hiss'.

The Taliban, having created this confusion, took full advantage by attacking the President's Palace, several of the army barracks and most of the police stations. By dawn, and with the surviving Taliban fighters making their escape, Kabul was in chaos. The President's Palace was a smoking ruin. The President's life had been saved by a daring rescue mission by an American helicopter which, amazingly, managed to complete its mission despite losing half of one of its tail-rotor blades. The timing of this whole operation had been carefully chosen by the Taliban to commence four hours after an optimistic speech by the American Secretary of State at the United Nations regarding the security situation in Afghanistan.

In Australia the next day, an authority on computers and a former lecturer on the subject at Sydney University was interviewed on State television. Demetri Emerton, though now in his early eighties, was still an impressive and intimidating figure. Most people who knew him well never dared to call him Demetri; it was always Dennis or Lofty. Even his lovely wife Ann, well aware of his real Christian name on her marriage lines, called him Dennis most of the time. When asked of his views on the development of the computer, Demetri urged caution. "The computer is a great aid to communication and the quest for knowledge, but we need to be careful in the information that we share." Emerton tapped the desk with his finger to make his point. "Computers do not think for themselves; they relay information that is available to them. But beware, the progress manufacturers are making is frightening; a computer that can think and reason is now within sight. Be warned."

These views on the future development of the computer were shared by many people, though few would offer the opinion on television. Without doubt events in Afghanistan and, recently, in Libya, suggested that a blind reliance on the system could be dangerous. In London, Peter Savage accompanied his Department Head to the morning's meeting. Ralph Maitland, a former Naval Captain, was furious over the shooting of a Navy Lieutenant Commander on the submarine HMS Astute and the resulting court case. "My God, Peter; what is the Service coming to when a young sailor, who is described as being immature and frequently in trouble due to his poor work performance, is allowed to serve on a submarine, of all places? It is beyond belief."

The room went quiet when Dame Helen Butters, the head of MI6, walked into the room. She looked around like a head teacher confronted by a difficult class. Her delicate nostrils twitched as cigarette smoke was detected. "Please put that disgusting object out; suck a sweet or something, at least it will freshen your breath." She looked across at Ralph Maitland and asked: "What progress on some bright juniors from that friend of yours in Penang?" Maitland nudged Savage. "Peter has the update on that, Boss." Savage took a breath. "I have not heard anything from Elizabeth Ascott-Gallop for at least a month. When we last spoke, she had two girls aged twenty who could be suitable. As cover, they would lodge at Cheyne Court under the protection of Carol Andrews–Dixon." "But that is a damned brothel," said the horrified head of MI6. Ralph Maitland spoke. "Not really, Boss; it is a club for influential people of both sexes, and it's a fount of information."

The meeting went on for another hour. Libya was now the main subject, with Savage fielding most of the questions. Dame Helen enquired if Savage thought the new Transitional Council was capable of controlling the rebel fighters once Gaddafi was defeated. Ralph Maitland gave Savage a sympathetic glance, very glad the question had not come his way. Savage took a breath, thinking this was no time to sit on the fence. "No, Dame Helen, I don't. Many of them will have old scores to settle and will do so." He paused and then went on: "In the last week I was in Libya, efforts to bring the various fighting groups into one formation to attack Sirte were having very little success and it is doubtful if the new leaders of

Libya can achieve this. The new regime is going to have a very difficult first few months."

That evening, Savage took Sally out for a meal and afterwards they spent the night together at a convenient hotel. In any organisation similar to MI6, married members of staff who are both in the Service are never permitted to work together. Both Sally and Peter Savage respected this rule and had decided that, for the time being, their relationship would remain secret. Over the following couple of weeks, several more incidents of computer sabotage came to light. One that caused some stifled laughter was at the Treasury, where a confidential programme on tax evasion was transferred to a daily newspaper instead of Scotland Yard. Furious Treasury officials tried to cover their embarrassment by suggesting Scotland Yard could have relayed the programme by accident. This claim was immediately dismissed by the grateful newspaper, which could prove that the document had come straight from the Treasury.

The Prime Minister made several speeches about the Arab Spring and Great Britain's role (meaning his) in freeing Libya from the tyrant Gaddafi. One speech that had the leaders of MI5 and MI6 glued to the television was from the United Nations, and was a plea from the Palestinians for Statehood after years of fruitless negotiations with the Israelis. Peter Savage had spent some time in that region and had friends in both camps and saw the biggest obstacle to a solution being the Israeli Prime Minister himself. What the American President would make of the Palestinians' request was anticipated to be America using its veto. With an election being just twelve months away, no candidate would throw away the influential Jewish vote or their generous financial help.

Chapter Forty

In China, senior secret service chiefs spent several hours watching film and reading reports from their many agents. Their successful use of 'Trojan Horse' insertions in many governments' computer systems had been a success on such a scale that had not been believed possible. Adding to the confusion had been the financial situation in many European countries, as well as America. There was also some unrest in many of the world's capitals. The street riots in Great Britain were largely triggered by the police shooting dead a shady local character and subsequently releasing an inaccurate statement of the circumstances that resulted in his death. At the end of the meeting it was decided to develop two more centres in Pakistan near the border with Afghanistan, which would produce programs of disruption inserted in memory sticks. These programs would be aimed at the financial centres, such as the banks and the world's stock markets.

Three days after the meeting with Dame Helen, Savage found himself in Cairo. The city that had suffered considerable damage during the overthrow of Mubarak was beginning to return to normal. Within a day of his arrival, Savage had managed to meet up with three shadowy figures from Egyptian Intelligence. Anxious to distance themselves from the previous regime, they and the rest of their colleagues had lain low during the violent days of the rebellion. But now, with the army back in control, they wasted little time in returning to their duties, making sure their masters were aware of their sworn loyalty to the new leaders.

Egypt was keeping a close eye on the way the Libyan situation was developing and had already toughened its stance towards Israel and resumed the movement of supplies to Gaza. Syria was another problem area, and the general feeling among the new Egyptian leaders was they preferred President Assad to remain in charge. Any concessions he chose to make would be his to live with. Savage was interested to find out if electronic sabotage had been a problem in Egypt; one of his contacts said "No," and the other two did not know what he was talking about. That night Savage took the contact he considered the brightest out for a meal. The man called himself Naza. It was not his real name, but the one he was known as in the murky world of intelligence.

During the meal, Savage asked if it would be possible for him to slip into Gaza without going through the Israeli check-points. Naza smiled. "That is not a problem, my friend; you could use the newly reopened supply route. The Israelis do not like it, but we no longer have our nose stuck up the Israeli arse as in the days of

Mubarack." The real purpose of Peter Savage's wish to get into Gaza was to find out if Hamas was attempting to slip a 'Trojan Horse' into the Israelis' highly-developed computer system. Once in Gaza, Savage was aware of at least three senior members of Hamas he could safely speak to. All three took money from British Intelligence which was discreetly transferred to their respective Maltese bank accounts.

Two nights later Savage was taken by Naza to the newly-opened border crossing. The area around the crossing was littered with bomb craters, evidence of the Israeli efforts to destroy the network of tunnels used by the people of Gaza to smuggle goods across from Egypt. A few words were exchanged between Naza and an Egyptian Army Sergeant, and Savage found himself sitting in the cab of a lorry filled with live chickens. The driver smiled a greeting and offered Savage a small cigar similar to the one he was smoking. Savage declined the kind offer; the stink from the lorry-load of chickens was preferable to that of the cigar, and less injurious to his health.

An hour later and Savage was dropped off near the centre of Gaza City. The driver and his load of chickens departed for the market area in a cloud of feathers and a strong smell of ammonia. Savage knew the Hamas headquarters were close to the city hospital, deliberately placed there hoping to avoid the Israeli bombing. They were to learn, of course, that the Israeli military considered that a target was a target irrespective of its sensitive location. The bomb-scarred walls of both buildings were clearly proof of this. At the Hamas entrance stood two armed guards, both scruffy and unshaven and clad in worn camouflage clothing. The weapons they carried, however, were spotless.

Savage mentioned a name to one of the guards, who shook his head. The second name produced a nod of acknowledgement and the guard took Savage into the building, after he had been briefly searched. They stopped at a door on the first floor and the guard knocked, then opened the door and spoke rapidly to the man inside. A face appeared and gave Savage a searching look then spoke to the guard, who walked away. "It seems a long time, Peter my friend, since I last saw you, but I am sorry to say little has changed here and the world seems to have forgotten us."

The two men talked generally for some minutes, and then Savage asked his question. The man shook his head, and then lit a cigarette, offering one to Savage, who declined. He took two puffs and blew out the resulting smoke, then said: "We are aware computers can be sabotaged and have in our organisation people who understand the technology, but to get into the Israeli system would be like breaching the security of Fort Knox." Savage spent the night in the building, sleeping on a pile of well-used blankets. In the morning he was given breakfast of strong, black coffee and half of a freshly-baked baguette. Half-way through the morning he was taken to see another agent, who questioned him further and asked if he could supply a memory stick containing a crippling virus which would disable any computer system.

"I don't doubt they exist," said Savage "Computers can and are protected against most know computer viruses, but a disruptive program known as a 'Trojan Horse' is a totally different matter." Savage thought about taking a chance and attempting to go into Israel, but thought better of it and decided to return to Egypt. He returned that evening in a very run-down coach full of men who were desperately looking for work. He decided to stay in Cairo for another two days and then arranged for a flight back to London. Ralph Maitland was disappointed that the trip to Cairo and Gaza had not produced any new information about the source of the disruptive memory sticks. China was the instigator of the sabotage, of that there was little doubt, but most of the agents passing on this technology were from Pakistan.

Savage was persuaded to pay a visit to GCHQ at Cheltenham. He met up again with Frank Grimley who, despite his expertise with computers, really would benefit from a hot bath. Grimley was amused to notice the twitching of the Savage nose as he was given a seat in the far-from-fragrant office. "I think their next target will be the financial sector," said Grimley "These so-called whiz-kids are highly intelligent but they lack common sense, and trust almost everything they see on their screens without question." Over the next half hour he demonstrated on his computer how several mythical companies could be set up, prompting interest from some traders anxious to invest some of their clients' money. Savage was fascinated by the nimble-fingered Grimley and stayed in the fetid office longer than he intended. A brisk walk to Cheltenham's railway station hopefully cleared his lungs of the 'Grimley odour', and he had to admit the trip had been worthwhile despite the possible threat to his health.

On his return to London he was disappointed to find Sally had been sent to Scotland with a computer engineer to check a problem that had been found on a submarine about to start a two-month patrol. He spent an hour with Ralph Maitland, who had been tasked by Dame Helen to decide if the time was right to disclose the location of the suspected memory stick programming factories to the American drone operators. Maitland asked Savage if he was satisfied the information was sound. Savage said it was. A flash signal was sent to the Combined Military Headquarters in Afghanistan with the map references of the targets, leaving them to decide what action to take.

When the information was received in Afghanistan, it tied in neatly with some intelligence extracted from a wounded Taliban fighter. The man, shot in the stomach and in great pain, was offered life-saving treatment in exchange for information. Anxious to live and return to his young wife, the man told all and, in doing so, not only recovered from his wounds but became a valued informant. At dawn the next day, three drones made two passes over their targets. In the first, they dropped leaflets giving the workers in the separate small factories five minutes to clear the buildings. On their second pass, they fired four rockets each into the now-deserted buildings, reducing them to a smoking heap of rubble. Whatever they had been producing was now lost forever, and the workforce had learned a valuable lesson.

The attitude of the Pakistani Government to this attack on several of its border villages was to complain, but not too loudly. The Pakistan Government was re-equipping its air force with used but updated American F16s and would certainly not wish to jeopardise that. The now-unemployed work-force was surprised to be offered Government work at the same rates of pay. Their work involved extracting a precious metal from tons of discarded mobile phones. The Chinese took these minor setbacks in their stride. Their targets now were the world's financial centres, where greed has replaced any trace of moral conscience and consequently this made them easier targets.

Sally Gunn returned from her trip to Scotland. They had found no evidence of deliberate sabotage, but the submarine's computer system would now be replaced. At lunch-time on the day following her return, she and Peter Savage took their sandwiches to an empty bench alongside the Thames, not far from the MI6 building. Cheese and tomato were not Savage's favourite fillings, but they were certainly Sally's. He quickly disposed of his snack and turned to the subject occupying his mind, which was marriage. The words that he had arranged neatly in his head to deliver refused to emerge with any fluency. He started, stopped, and then started again, but the words just would not appear in the right sequence.

Finally, Sally took his hand and pressed it gently to her lips, then said: "Peter, will you marry me and be the father of my children? I am willing to go back to my police career but, if you hurry up and get me pregnant, I won't have to." They embraced, no words of acceptance being necessary. A passing policeman paused and considered if they were in breach of any local by-law. Having decided they were not, he continued his beat. When they returned to MI6 after lunch, Madge was the first to hear of their news; then Savage went to see Ralph Maitland to make his intention official. The old sea-dog was delighted and wished them well, then mentioned that Sally would have to work on separate cases or return to the police.

The pair married two months later, and spent an idyllic week in the Isles of Scotland. One evening, as they ate supper in their rented cottage, the radio music hour was interrupted to announce a worldwide stock-market crash and the failure of two banks that were household names. Savage switched off the provider of the bad news, saying: "This week it is not our problem; next week it may be but, here and now, the world is just about you and me."

The End

Other books by the same author

Oath of Allegiance

Strange Alliance

Jungle Haven

The Drums of War

Crown Assassin

The Warriors

The Savage Years

The Satan Seed

Southern Adventure

Exodus Five

The Shadow Man

The Final Solution

The Last Sunrise

9781780352602
ND - #0188 - 110926 - C0 - 197/132/28 - PB - 9781780352602 - Gloss Lamination